THE
DIAMOND
MAKERS

Humans have treasured diamonds for their exquisite beauty and unrivaled hardness for thousands of years. Deep within the earth, diamonds grow. Diamonds the size of footballs, the size of watermelons – billions of tons of diamonds wait for eternity a hundred miles beyond our reach. But how is it possible to create these spectacular gems at the earth's surface?

Spanning centuries of ground-breaking science, bitter rivalry, outright fraud, and self-delusion, *The Diamond Makers* is a compelling narrative centered around the brilliant, often eccentric, and controversial pioneers of high-pressure research. This vivid blend of dramatic personal stories and extraordinary scientific advances – and devastating failures – brings alive the quest to create diamond. Scientists have harnessed crushing pressures and scorching temperatures to transform almost any carbon-rich material, from road tar to peanut butter, into the most prized of gems. The book reveals the human dimensions of research – the competition, bravery, jealousy, teamwork, and greed that ultimately led to today's billion-dollar diamond synthesis industry.

ROBERT M. HAZEN is a research scientist at the Carnegie Institution of Washington's Geophysical Laboratory and Robinson Professor of Earth Sciences at George Mason University. He is a graduate of MIT and Harvard and a leader in the scientific literacy movement. Hazen is the author of more than 200 articles and 14 books on science, history and music, and appears frequently on radio and television programs on science. His books have achieved widespread critical acclaim, and include *The Breakthrough: The Race for the Superconductor, Science Matters: Achieving Science Literacy* (co-authored with James Trefil), *Keepers of the Flame* (with M. H. Hazen), and *Why Aren't Black Holes Black?* (with Maxine Singer). He is also a professional trumpeter, and has performed with the New York City Opera, the Royal, Bolshoi, and Kirov Ballets, the Boston and National Symphonies, and the Orchestre de Paris. He lives in Glen Echo, Maryland, with his wife and two children.

THE

DIAMOND

MAKERS

ROBERT M. HAZEN

CAMBRIDGE
UNIVERSITY PRESS

PUBLISHED BY THE PRESS SYNDICATE OF THE UNIVERSITY OF CAMBRIDGE
The Pitt Building, Trumpington Street, Cambridge CB2 1RP, United Kingdom

CAMBRIDGE UNIVERSITY PRESS
The Edinburgh Building, Cambridge CB2 2RU, UK   http://www.cup.cam.ac.uk
40 West 20th Street, New York, NY 10011–4211, USA   http://www.cup.org
10 Stamford Road, Oakleigh, Melbourne 3166, Australia

First published 1999

Typeface in FF Scala 10.3/12.85pt *System* QuarkXPress®   [SE]

A catalogue record for this book is available from the British Library

Library of Congress cataloguing in publication data

Hazen, Robert M., 1948–
The diamond makers / Robert M. Hazen.
p.   cm.
Includes index.
ISBN 0 521 65474 2 paperback
1. Diamonds, Artificial.   2. High pressure (Technology)
I. Title.
TP873.5.D5H39   1999
666′.88–dc21   98–49423   CIP

ISBN 0 521 65474 2 paperback

Transferred to digital printing 2003

CONTENTS

PREFACE

ON A COLD WINTER DAY IN DECEMBER 1955, Robert Wentorf Jr. walked down to the local food co-op in Niskayuna, New York, and bought a jar of his favorite crunchy peanut butter. This was no ordinary shopping trip, for Wentorf was about to perform an experiment of unsurpassed flamboyance and good humor. Back at his nearby General Electric lab he scooped out a spoonful, subjected it to crushing pressures and searing heat, and accomplished the ultimate culinary *tour de force*: he transformed peanut butter into tiny crystals of diamond.

The triumph of Wentorf and his GE colleagues, who learned to turn just about any carbon-rich material, from road tar to snake oil, into a sparkling treasure, marked the culmination of centuries of scientific study and technological breakthroughs – research driven by a fascination with nature's most romantic gemstone and by our awe at the power of high pressure to transform matter.

Scientific research encompasses much more than the dry data and abstract equations that appear in journals and monographs. Science is a marvelously human endeavor, driven by curiosity and ambition and wonder and greed. *The Diamond Makers* tells the story of scientists who spent their lives enthralled with the transforming power of high pressure. Under pressure, liquid water crystallizes into new kinds of ice, everyday gases condense to metals, and black soot becomes a precious gemstone. The science and technology of diamond making pervades our society, from one-hour eyeglasses to efficient road repair. Synthetic gem diamonds also play a central role in the ongoing adventure of high-pressure research, as researchers squeeze together pairs of diamonds to produce pressures that exceed those at the center of the Earth.

The synthetic diamond story spans more than a century of top-notch science and outright fraud, brilliant insight and self delusion. In spite of the physical risks and economic rewards, most of the high-pressure pioneers were driven ultimately by an overpowering desire to explore the unknown. *The Diamond Makers* is the story of scientists and engineers and the powerful machines they have built to unlock Nature's secrets.

* * *

Every new scientific theory must be considered in the social and intellectual context of its time. Scientists find it tempting to describe the history of science as a series of successes and failures, judged against our modern "correct" theories. Such an approach might seem especially applicable to the diamond story, for there is one simple measure of success: was diamond made in the laboratory? But such a narrow perspective is at best misleading. Science builds on past experience, and discovering what doesn't work is often the key to finding what does. Those dedicated scientists who stumbled and fell short of our perception of truth cannot be viewed as failures, and their story deserves to be told.

Some philosophers criticize scientists who believe in an independent, knowable Nature. Reality, they say, is merely a human construct: There is no absolute truth in nature, only our inadequate perception of that truth. Some scholars even deny the concept of "progress;" for them there is only change. I disagree. Along with most scientists, I am an unabashed believer in a universe with knowable physical laws, and I believe that we can measure scientific progress by our degree of understanding of those laws. To be sure, we have only come the tiniest fraction of the way toward understanding, but each advance marks a triumph for science and technology. The story of the diamond makers represents one such triumph. A few decades ago no one had ever done it; now, anyone with a few hundred thousand dollars and a big basement can make them at home. To me, that's progress.

This book is not a comprehensive account of the history of diamond synthesis. Rather, I want to share some of the lore of science and scientists – the stories that never make it into formal publications. I was drawn into high-pressure research by these remarkable people. I have shared the drama of their discoveries, as well as the frustration of their failures.

Most of this book focuses on the struggles of scientists who first attempted to make diamonds in the laboratory. The history of diamond synthesis cannot be divided neatly into pure versus applied stages of research, as is the case with many of the scientific discoveries that have transformed the twentieth century. With new diamond-making technologies have come both exciting new ways to study matter and extraordinary new uses for synthetic diamonds. As diamond making has become routine, the gems themselves have allowed scientists to achieve higher and higher pressures.

I have tried to present a fair and balanced history of the diamond story, but I cannot escape biases developed during more than 20 years at the Carnegie Institution of Washington's Geophysical Laboratory, home of many central players in the high-pressure game. I am more intimately familiar with work done in the United States than elsewhere, so most of the book focuses on American scientists. Key contributions by researchers in Europe, the former Soviet Union, Japan, and Australia have perhaps received less than their fair share of space; I encourage workers in those regions to recount their own personal histories.

* * *

I wrote this book for three reasons. First, diamond has captivated human imagination for thousands of years, and the fascinating story of the gem's synthesis is not widely known outside of the technical literature. Second, diamond making provides an ideal framework for describing high-pressure science – a field that has profoundly shaped our understanding of the material world and enhanced our ability to alter it. But most important, this book arose as an answer to a rhetorical question posed by my son in the winter of 1990. After a particularly frustrating textbook lesson in his junior high school science course, he asked me "Why would anyone want to be a scientist?"

This book may be a longer answer than Ben expected, but I hope it will help him to see the drama that underlies scientific discovery. I hope it's not too late for him to appreciate how powerful science can be in helping us to understand our world, and, if used wisely, to shape it to our benefit.

* * *

The greatest joy in writing this book has been the opportunity to learn from so many high-pressure pioneers. Sadly, Percy Bridgman died before my years in Cambridge, Massachusetts, but many of his students and younger associates have vivid memories of his unique style and his extraordinary laboratory. Hatten S. Yoder, Jr., Alvin Van Valkenburg, Francis Bundy, Herbert Strong, Robert Wentorf Jr., and Tracy Hall all met Bridgman in his later years, and their recollections have enriched the story. I have also relied heavily on Bridgman's recent biography, *Science and Cultural Crisis* by Maila L. Walter (Stanford University Press, 1990) for biographical details.

Erik G. Lundblad, a member of the Swedish electric firm ASEA's diamond-making team, contributed much information on Baltzar von Platen's personality and inventions, as well as on ASEA's QUINTUS project. Lundblad also provided fascinating historical photographs from the ASEA effort.

Loring Coes, Jr., was remembered by many of his long-time associates at Norton: Neil Ault, George Comstock, Paul Keat, Alan G. King, and Osgood Whittemore all shared their reminiscences. Additional information was provided by Francis Birch, Francis R. Boyd, Eugene C. Robertson, Alvin Van Valkenburg, and Hatten S. Yoder, Jr., who were at the Norton meeting on December 4, 1953, and Edward Chao, who discovered coesite in nature. Samuel Coes graciously provided details on his brother's early years.

All of the original members of the General Electric diamond-making team have contributed significantly to this book. Francis Bundy, Herbert Strong, and Robert Wentorf in Schenectady, New York, spent many hours retelling the story, clarifying details, and showing the site where the "Man-Made Diamond" process originated. Tracy Hall in Provo, Utah, also shared his memories of four decades of diamond making, and patiently explained the nature and origins of the disagreements that have arisen regarding the GE history. Harold Bovenkerk, the last member of the original team to retire from General Electric, provided his unique insight regarding the history of domestic and foreign diamond synthesis. All of these men were generous in sharing historic photographs and documents, as well as unpublished anecdotes. I am also grateful to Mark Sneeringer and Anne Shayeson who hosted a memorable tour of General Electric's Worthington, Ohio, diamond making facility.

Francis R. Boyd, Ivan Getting, Julian Goldsmith, and Alvin Van Valkenburg, along with all the original GE diamond makers, provided stories about George Kennedy's extraordinary life, scientific research, and diamond synthesis endeavors. Armando Giardini contributed much on the unpublished history of diamond making at the Army

Electronics Command, Fort Monmouth, New Jersey. Henry Dyer of De Beers Industrial Diamond Division, and one of the original De Beers diamond synthesis team, shared the history of South African synthetic diamond studies. O. L. Bergmann of Du Pont shared information on Mypolex, an explosively manufactured diamond.

Alvin Van Valkenburg, inventor of the lever-arm diamond anvil cell, had been my friend for almost two decades. Shortly before his death in December 1991, he spent many hours recounting the high-pressure history he knew so well. His son, Eric Van Valkenburg, contributed important photographs and other documents. Bill Bassett, who with his colleague Taro Takahashi learned about the diamond cell from Van and thus became the first earth scientist to use it, shared many memories of his research and his many friends and colleagues in high-pressure research.

Most of all, I am indebted to my many present and former colleagues at the Carnegie Institution of Washington's Geophysical Laboratory. For almost 90 years, the Geophysical Laboratory has played a leading role in high-pressure research, especially in the earth sciences. Hatten S. Yoder Jr., whose work in high-pressure began more than a half century ago, continues to provide inspiration and expertise to the next generation of diamond makers. This book has been shaped by his wealth of first-hand historical knowledge.

Francis R. (Joe) Boyd, who could have been the first diamond maker when he came to the Geophysical Lab in 1953, is still making big presses and squeezing rocks. Joe's decades of research on diamond-bearing rocks from South Africa and elsewhere was critical in telling the story.

I owe a special debt to Peter M. Bell, who advised me in my first years at the Geophysical Lab, and his brilliant colleague Ho-Kwang (David) Mao, who has transformed modern high-pressure research. Peter and David were the first to achieve a million atmospheres pressure in a diamond cell, and David, either directly or through his students, has instructed most of the handful of humans who have learned to duplicate the feat. Now, with Russell J. Hemley and a gifted team of junior colleagues, David Mao continues to make high-pressure history. All of these scientists have contributed immeasurably to my historical research.

I have received thoughtful and constructive reviews of the manuscript from many friends and colleagues. Those who have read and reviewed substantial portions of the manuscript include Allen Bassett,

William Bassett, Harold Bovenkerk, Ray Bowers, Francis Boyd, Francis Bundy, James Cheney, Tracy Hall, Russell Hemley, Raymond Jeanloz, David Mao, Charles Meade, Charles Prewitt, Gretchen Prewitt, Herbert Strong, Robert Wentorf, and Hatten Yoder, Jr. Each of these high-pressure experts made important additions and corrections, and to each I am indebted.

Finally I thank Margaret H. Hazen, my greatest supporter and most perceptive critic, whose influence is present on every page of this book.

PROLOGUE

❖

PRESSURE. TO MOST PEOPLE THE WORD brings to mind the daily stress of our busy lives. Yet to many scientists, pressure means something very different; it is an idea filled with wonder and power – a phenomenon unlike anything else we know. Pressure shapes the stars and planets, forges the continents and oceans, and influences our lives every moment of every day.

Three arenas of high technology drive our economy and propel us toward the twenty-first century. Information technologies link us to the world and its resources as never before, while biotechnologies hold the tantalizing promise of an era of health and plenty for all. But of all the advances that shape this age, none plays a more immediate and dramatic role in our day-to-day lives than the science and technology of new materials, and no technology holds more promise for creating and producing these remarkable new substances than the technology of high pressure. With pressure we can transform liquids into spectacular crystals, ordinary gases into exotic dense metals, and lumps of coal into precious gems.

We experience pressure whenever we push a button, press down with a pen, pump up a tire, or hear the explosive birth of popcorn. Behind such simple actions as inflating a party balloon or squeezing a tube of toothpaste lies a phenomenon that has captivated scientists, driven engineers, and transformed virtually every aspect of our physical world. Pressure occurs whenever a force acts on an area. Your shoe applies pressure to the sidewalk, water exerts pressure on a skin-diver, and the atmosphere weighs down on everyone and everything at the earth's surface with a pressure of about fourteen-and-a-half pounds per square inch. Scientists call normal, everyday pressure – the weight of thirty or forty miles of air pushing down on you – one atmosphere or one bar.

Most of the pressures we experience in everyday life are rather modest. A pressure cooker generates about one-and-a-half atmospheres, while the air in your car's tires is typically pressurized to about two atmospheres. A scuba diver 300 feet down experiences almost ten atmospheres of pressure, while a heavy-set woman in stiletto heels could apply a pressure of about fifty atmospheres to the floor on which she walks.

However, when researchers talk about high pressure, they mean vastly greater pressures of thousands or millions of atmospheres – kilobars or megabars. In fact, as pressure records have soared during the past three decades, new superlatives have come into play. "Superhigh pressures" (hundreds of thousands of atmospheres) and "ultrahigh pressures" (millions of atmospheres) are now standard jargon.

Scientists investigate these immense, often dangerous extremes because pressure causes matter to change in remarkable ways – much like the dramatic changes induced by temperature. Humans have known for thousands of years that liquid water freezes to a solid if cooled and boils to a gas if heated. We can now mimic these so-called changes of state at room temperature by using pressure: we can cause water to "freeze" at high pressure and turn it into gas in a vacuum.

Over the centuries humans have learned to use a wide range of temperatures, from a tiny fraction of a degree above absolute zero (the coldest possible temperature) to about 1 000 000°C at the focus of several intense laser beams. Over this temperature range we observe remarkable changes in matter, from superconductivity to nuclear fusion. During the past few decades, scientists have learned to exploit a similar range of pressure, from a high vacuum – less than a billionth of an atmosphere – to pressures of many millions of atmospheres. Matter subjected to these extremes of pressure displays astonishing behavior, rivaling anything seen at exotic temperatures. Researchers have discovered that high pressure can turn ordinary compounds into superhard abrasives and transform everyday rocks and minerals into the dense materials that form the dynamic interior of our planet.

Pressure alters matter by forcing atoms into ever smaller volumes. Every substance displays this phenomenon: as pressure increases, the volume of the compressed material decreases. At high pressures atoms must shift into more efficient, more densely packed arrangements. At high enough pressure any gas will become a solid, and every solid will adopt a new, denser form. To accomplish this volume reduction, the bonds between atoms – the interaction of the atoms' electrons – must also change. Pressure thus serves as a powerful probe of atoms and their electronic structure, helping us to learn how our physical world holds itself together.

CHAPTER I

MYSTERIES

The most highly valued of human possessions, let alone gemstones, is the "adamas," which for long was known only to kings, and to very few of them.
PLINY THE ELDER, *Natural History*, Book XXXVII

DEEP WITHIN THE EARTH DIAMONDS GROW. Diamonds the size of footballs, diamonds the size of watermelons – countless billions of tons of diamonds wait for eternity a hundred miles beyond our reach.[1]

Thousands of years before the invention of science, humans treasured the glistening, hard stones they found among the gravels of exotic rivers, without knowing exactly what they were or how they came to be. Some said diamonds were pieces of stars fallen to earth, or perhaps the remains of water frozen for too long, while others spoke of crystals grown at ocean depths or formed in the path of lightning bolts. Despite people's curiosity, the true origin of diamonds remained a mystery until early in this century.

Humans have always prized these bits of matter for their unrivaled physical properties.[2] They sparkle as the most brilliant of gems, dispersing light better than any other precious stone. Diamonds last almost forever, withstanding the most corrosive salts and acids for aeons. Pure diamonds provide superb electrical insulators, while they conduct heat energy more efficiently than any other substance. They boast exceptional chemical purity, often more than 99.9% carbon. And, of course, diamond is the hardest known material – almost twice as hard as any other natural or synthetic creation.

Ancient artisans recognized at least some of diamond's unique characteristics, and were probably the first to study the stones' physical properties. If struck forcefully, they found, diamonds could be broken into chips and shards of great hardness and utility. Diamond-edged knives and engraving tools have been used by stone workers for millennia, and some ancient warriors are thought to have embedded diamonds in their edged metal weapons.

Yet as useful as they may be, diamonds have long been coveted as much more than mere physical curiosities. They are the stuff of magic and legends. Mystics and alchemists ascribed wondrous attributes to the stones, which were said to grant the wearer awesome strength on the battlefield, as well as potency in the bedroom. Wise men proclaimed diamond to be a protection against evil, an antidote to poison, a cure for insanity, and a charm for women in childbirth.

Flamboyant stones possessing unusual size and quality have long had a special mystique.[3] Large diamonds weighing a hundred carats or more (a carat being a fifth of a gram, or about the weight of a single small pea) traditionally are given glamorous names, and some notable stones have even developed reputations. The magnificent Orloff diamond, a 190–carat gem that adorns the imperial Russian scepter, was stolen by a French deserter from the Indian foreign legion, in a sacrilegious exploit that has served as a prototype for countless dime-novel and adventure-movie plots. He disguised himself as a devout worshiper and plucked the gem from the eye socket of an idol of the god Sri-Ranga. The stone was smuggled to France, and there purchased by Russian Prince Orloff as a tribute to Catherine the Great.

The breathtaking, deep-blue Hope Diamond, now the most popular exhibit of the Smithsonian Institution in Washington, D.C., is said to carry its own curse. Brought to Europe from India in the seventeenth century (some say it, too, was stolen from the eye of a vengeful idol god), the striking 112–carat stone from which the Hope was cut had many unfortunate owners. Louis XIV of France, Marie Antoinette, Queen Maria Louisa of Spain, and a sad sequence of wealthy, ill-fated Americans all possessed and wore the gem before New York diamond merchant Harry Winston acquired the Hope and sent it to the Smithsonian's National Museum of Natural History.

To these well-known stories can be added dozens more. The 410–carat Regent Diamond was reputedly smuggled to Europe by an Indian slave who sliced open his own leg and buried the diamond deep in the wound. The great stone adorned the hilt of Napoleon's sword and the crown of Charles X, and now serves as the centerpiece of the French Crown Jewels at the Louvre. The Grand Sancy Diamond, also at the Louvre, became a favorite turban adornment of the prematurely balding King Henry III of France. Striking colored diamonds – the Chantilly Pink, the Dresden Blue, the Florentine yellow, the Tiffany golden diamond, and dozens more all have their stories. And then there are the giant stones – the 726–carat Vargas with which a slave woman won her freedom in 1850 Brazil, the 995–carat Excelsior found in a shovel full of gravel at Jagersfontein in South Africa, and of course the Premier Mine's fabulous 3106–carat Cullinan, the largest gem diamond ever found.

Late in the afternoon on January 2, 1905, a mineworker alerted the Surface Manager, Fred Wells, to a brilliantly shiny object that caught the setting sunlight high on the wall of the diggings. Wells carefully inched his way to the spot and used his pocket knife to pry out the colossal stone; his understated reaction, according to a bystander, was that the mine's principal owner Colonel Thomas Cullinan, "will be pleased when he sees this!" Wells hurried to the mine office to have the epic find weighed, but the office staff was unimpressed. "This is no diamond," the inspector said, and threw the stone out the window.

Wells went back outside to retrieve the diamond (fig. 1), which was eventually authenticated and logged in at more than one-and-a-third pounds – more than three times larger than any other known diamond. Upon examining the mass, awestruck geologists realized that the Cullinan represented just a small piece from what must have been a much larger, eight-sided crystal. For two years the giant diamond was

Fig. 1 The magnificent 3106-carat Cullinan diamond, the largest ever found, was at first tossed aside by a mine official who thought it a worthless piece of rock. The stone's discoverer, Fred Wells (right), stands with the Premier Mine manager William McHardy (center) and owner Thomas Cullinan. (Courtesy of F. R. Boyd.)

exhibited to prospective buyers at a London bank in its natural state; in early 1908 the stone was cleaved to yield nine principal faceted gems. The two largest faceted diamonds in the world, both cut from the Cullinan, now grace the Imperial Sceptre and State Crown of Great Britain.

* * *

As fascinating as the human histories of individual gemstones may be, they pale beside the natural saga of diamond's unimaginably ancient birth and improbable epic journey to human hands. The realization of diamond's superlative properties and the discovery of its violent origins frame a marvelous scientific adventure.

Today, most people admire diamonds for two exceptional attributes: their hardness and their brilliance. Scholars knew of diamonds' unrivaled hardness since antiquity, but its unique optical properties went unrecognized until comparatively recently. The natural diamonds that were hoarded by potentates of old have little of the visual drama that we associate with today's faceted gems. Deeply colored rubies, emeralds, and sapphires were far more prized as adornments. Owners accumulated the seemingly indestructible diamond pebbles, as found in their unpolished natural state, as talismans against defeat and symbols of their own "manly" virtues, without ever seeing diamonds as objects of beauty.

When unearthed, most diamonds appear roughly rounded with perhaps a hint of regular crystal form. Many are colorless, but most are pale shades of yellow; red, orange, green, blue, brown, and even black diamonds are also found. Raw, uncut stones lack the exuberance of jewelry store gems, and can appear quite ordinary; Brazilian gold miners of the eighteenth century cast aside a fortune in unrecognized diamonds while panning for the precious metal. Diamond's familiar ornamental role represents a relatively recent development – a consequence in part of scientists' growing understanding of the nature of light.

A few scientific ideas have become part of our folklore. The equation $E = mc^2$ is one – a cultural icon as much as it is a statement of the equivalence of mass and energy. Another of these commonplace science snippets tells us that the speed of light is a constant. A T-shirt slogan popular in physics departments proclaims "186 000 miles per second:

It's not just a good idea, it's the law!" But as for many other legal systems, there's some fine print most people ignore. You have to add the rather mundane words, "in a vacuum." When light travels through matter – air, water, glass, or diamond, for example – it travels slower than 186 000 miles per second. The actual explanation, having to do with the way light interacts with the electrons present in every atom, is somewhat complex, but you can visualize this slow-down by thinking of light rays having to make little detours every time an electron gets in the way.[4]

Most clear and colorless objects retard light only a modest amount. The air we breathe has only a trifling billion-trillion atoms per cubic inch. Spaces between atoms are much greater than the size of the atoms themselves, so air reduces light speed by just a few hundred miles per second – not enough to notice under most circumstances. In water and ice, which have thousands of times more atoms per cubic inch than air, light travels about 140 000 miles per second – 30% slower than in a vacuum. Window glass drops light speed to 120 000 miles per second, similar to the travel time through most common minerals, whereas lead-containing decorative glass, the kind used in chandeliers and cut glass, slows light even more, to about 100 000 miles per second (lead has lots of electrons that get in the way). Diamonds put the brakes on light like no other known colorless substance. Diamond is crammed with electrons – no substance you have ever seen has atoms more densely packed – so light pokes along at less than 80 000 miles per second. That's more than 100 000 miles per second *slower* than in air.

Most people never have reason to notice the variable speed of light, but you experience one of its consequences every day. Each time light passes from one clear substance into another with a different light speed, the light rays have a tendency to bend. You've probably noticed the distortion of people and objects in a swimming pool, which occurs when light waves have to change direction as they pick up speed coming out of the water. Ripples on the pool's surface compound the angular distortion. If you wear glasses or contact lenses, which "correct" the way light bends into your eyes, you take advantage of this useful optical phenomenon.

Light does not always bend when passing between different materials. If light rays strike a clear substance head on or at a modest angle – like the path of light coming through your window – most of the rays will travel straight through without bending. You can look down from a boat at the nearly undistorted bottom of a calm, clear lake or pond because sunlight enters the water from overhead and then comes back through the transparent water almost vertically to your eyes. But try as you might, you can't see the bottom of even the clearest lake standing on the shore, because you are at too low an angle to the water. Almost all of the light reaching your eyes has been reflected off the water's surface. That's why you can see the beautiful mirrored reflection of trees on the opposite shore of a glassy lake early in the morning.

Diamond plays this reflecting trick better than any other colorless substance. Light enters a faceted gemstone from all sides, but it may bounce back and forth several times inside before it finds a clean, straight shot out. All this changing direction accomplishes something very dramatic, because so-called white light actually contains all of the rainbow's colors. Each color – red, orange, yellow, green, blue, and violet – bends and reflects inside the diamond slightly differently. The farther the light travels, the more the colors separate, or "disperse." Bounce light inside a diamond just two or three times and the colors disperse spectacularly. Diamond-like substitutes, including "cubic zirconia," a crystalline compound synthesized from the elements zirconium and oxygen, attempt to mimic this light-dispersing property, though they fall short of diamond's brilliance and unrivaled hardness.

If you look closely at a faceted diamond, you can see that it soaks up white light and breaks it apart like a prism, dispersing it into a rainbow of colors. Diamonds sparkle and dance with colored light; each of its dozens of facets produces its own dazzling display. Other natural gemstones disperse white light to some degree, but none comes close to diamond's ability to reveal the rainbow.

No one can say for sure when faces were first cut and polished on a diamond, but we are certain that the process was extremely tedious.[5] Only diamond powder possesses the hardness to polish diamond facets effectively, and achieving a smooth surface by hand is a lengthy and exacting task. Diamond cutters, who cleave and polish natural stones, are thought to have been active in India a thousand years ago, while Parisian documents record such artisans working in France as early as the fourteenth century. The original motivation for imposing flat faces on rough and rounded stones was probably simply to remove unsightly impurities. But, perhaps quite by accident, diamond finishers found that faceting accomplished much more than simple cleaning. Faces enhanced the beauty (and value) of their pebbles, and transformed human perceptions of diamond.

The shaping of a raw diamond into a pleasing geometric form represents a painstaking and nerve-racking process. The diamond cutter first examines the stone for cracks and impurities, then designs a strategy for extracting the largest possible gem. Every diamond possesses some internal planes that are slightly weaker than others – cleavage planes that the diamond cutter hopes to find and exploit in shaping a stone. To cleave a gem diamond, the craftsman first uses a diamond-tipped instrument to inscribe a slight groove on the outside of the gem along the desired splitting direction. The cutter then places a sharp chisel on the groove and strikes the chisel with a swift, sharp hammer blow. Diamond cutters do not take their task lightly, for there is always a chance that a diamond, if improperly "read" or poorly struck, will shatter into small pieces. The largest diamonds often require years of planning before cleaving is attempted. Rough stones can also be sawed with a thin wheel coated by diamond dust and oil, but this tedious process requires many hours for even the smallest stones.

Once a gem has been roughly shaped by cleaving or sawing, the slow process of polishing facets begins. A metal plate, its surface impregnated with diamond dust, spins at high speed while the diamond is precisely oriented for the creation of each of its many faces. In ancient times, cutters arranged their faces haphazardly, with little effort to compute the most favorable angles to achieve maximum light dispersion and a pleasing symmetrical cut. The oldest known cut stones, recorded and illustrated by the French traveler and pioneer diamond merchant Jean Baptiste Tavernier in the mid-seventeenth century, appear crude and ungainly compared to modern jewels. But even ill-cut facets made diamonds more attractive, and people began to prize the gems for being beautiful as well as durable.

Popular history records that in the 1450s King Charles VII of France became the first European to give a gift of diamond jewelry to a woman, the beautiful Agnes Sorel. That the woman was his mistress, not his wife, may have delayed the adoption of the custom of the diamond engagement ring, but plenty of women, wives and mistresses alike, proved willing and eager to display the new symbols of wealth and fashion. In a twinkling, the gemstone of choice for powerful men became the adornment of choice for powerful women.

Demand for diamonds soared and, for more than two centuries,

India successfully met that demand. European diamond cutters sought to create ever more striking diamond jewelry by devising new ways to cut and polish the brilliant gems. Symmetrically faceted stones, championed by the Belgian craftsman Louis de Berquem, became the norm by 1500. Others added extra facets and introduced fancy new shapes – emerald cut, pear shape, marquise, and the popular 57–facet brilliant cut – to the diamond cutter's repertoire. More facets, they found, produced a more dramatic play of colors, while a variety of pleasing shapes let them make the best use of irregular stones. Carefully chosen angles between flat faces – first discovered by trial and error, but eventually calculated with mathematical precision – increased a diamond's ability to disperse white light into colors and thus enhanced its beauty even more. Master diamond cutters learned the secrets of sculpting magnificent gems from crude, irregular pebbles. Spectacular rings, bracelets, necklaces, and pins gave the very rich a new kind of bauble on which to spend their money – a new way to flaunt their wealth.

Making and selling diamond jewelry became a thriving business, so much so that by 1700 diamond demand had outpaced the dwindling Indian output from the almost exhausted alluvial mines. But, as luck would have it, miners discovered a major new source of diamonds in the Portuguese colony of Brazil in 1726. The diamond frenzy of the very rich continued unabated for another century.

Today, of course, the story has changed; now, almost everyone can own a diamond. Through a combination of new large-scale mining centers and brilliantly effective advertising, the once exclusive gemstone has become the symbol of love for people throughout the world. Today most people buy diamonds simply for their exceptional beauty, but the gem's particular mystique has not been lost. Diamonds are still sold as talismans by advocates of "crystal power," while seductive TV models look us in the eye and whisper, "Diamonds are love!" as though the hard, sparkling chips might somehow hold the power to alter our destinies.

* * *

By the seventeenth century the Western world had learned of diamonds' strength and beauty, but great mysteries remained unsolved. Two puzzles framed the subsequent centuries of diamond research: Of what are diamonds made, and how do they form in nature?

The first clues came from researchers who subjected diamonds to light and heat. Sir Isaac Newton marveled at the unique optical properties of diamonds during his attempts to understand the nature of the sun's radiation.[6] The gem's well-known ability to break white light into rainbow colors – a trick diamond performed better than any other substance known at the time – provided Newton with a challenging puzzle. Assuming that oils possess the greatest light-dispersing powers, Newton speculated that diamond was "probably an unctuous substance coagulated" – a solidified oil. He also came to a startling conclusion – diamond, like oil, would be combustible.

Newton's speculation resolved neither of the two central diamond questions, nor were any of his contemporaries able to push any closer to the answers. A century later, scientists tackled the diamond mystery from a new perspective, using the young science of chemistry. All matter, chemists had found, forms from a few different kinds of building blocks, which they called atoms. Chemists had embraced the task of learning which combination of atoms forms each fragment of the material universe, and they approached their mission like destructively curious architects who rip apart buildings brick by brick, plank by plank, to identify the materials used in their construction. Tearing matter apart also gave them hints about how to put it back together – the chemist's second mission.

Chemists devised many new ways of analyzing their world, but their first, best tool was fire. Many materials burn, leaving behind an ash residue that researchers could identify using simple chemical tests. And yet, despite Newton's conjecture about the flammability of diamond, attempts to ignite the hardest known substance seemed futile. Ordinary flame didn't work, nor did most ovens, but chemists kept trying. Finally, in 1772, the brilliant French chemist Antoine-Laurent Lavoisier confirmed Newton's prediction when he focused sunlight onto a diamond.[7]

Lavoisier's private laboratory at his estate near Blois was a far cry from the high-tech chemistry facilities of modern research. He performed his experiments in elegant surroundings; his lab was filled with fine balances and microscopes and other instruments crafted of rich mahogany and polished brass. Hand-blown bottles with thick cork stoppers lined his shelves, and a sturdy wooden workbench supported an array of exotic beakers, burners, and other chemical apparatus. Lavoisier, resplendent in powdered wig and fine embroidered coat, conducted his experiments with great care while his wife, Marie, recorded the results.

The chemist prepared his diamond specimen by sealing it in a glass jar filled with oxygen. He selected a thick convex glass lens that concentrated the sun's heat many fold, and mounted the lens on a pivoting metal stand that would keep the light focused on the gem as the sun moved. Lavoisier watched the diamond glow and burn like a piece of charcoal, not just to ash, but completely and totally away. The only by-product, identified in the laboratory as the gas carbon dioxide, suggested a close chemical relationship between diamond and charcoal, which also produces carbon dioxide when burned. Unfortunately, Lavoisier did not pursue this intriguing discovery, turning instead to other chemical studies in a distinguished career tragically cut short by the guillotine of the French Revolution.

The English chemist Smithson Tennant expanded on Lavoisier's work, demonstrating conclusively that diamond is nearly pure carbon, differing from charcoal only in its external form.[8] By converting identical weights of charcoal and diamond to exactly the same volume of carbon dioxide gas, he established the chemical equivalence of the two dissimilar solids. Tennant's results, published in 1797, astonished the scientific world. How could it be that the toughest material on earth forms from pure and simple carbon, the element of coal and soot? Scientists knew that carbon usually forms graphite, a material so soft that it is valued as a lubricant, so black that it is employed in the finest pencil leads. Many scientists, especially those who suspected that diamond must conceal a second, as yet unidentified element, were reluctant to believe Tennant's conclusions. Not until two decades later, after many careful researchers duplicated Tennant's observations, were these results accepted by the scientific community at large. Smithson Tennant, like Lavoisier before him, had little chance to enjoy this vindication; in February of 1815, while examining fortifications in Napoleonic France, he was killed in a bizarre accident as his horse fell through a decaying drawbridge.

Carbon taunted the chemists. With the exception of a few obscure diamond-bearing streambeds in India and Brazil, carbon occurred as graphite. Why did nature produce two kinds of carbon? Could scientists reproduce nature's feat? For more than a century they tried and failed. Physicists and chemists used every trick they could think of to grow the gems. Some attempted to form diamond by evaporating carbon-rich solutions. Others tried electric arcs or intense heat to coax the carbon atoms into their most desirable form.

Nothing worked.

* * *

The discovery of primary diamonds – precious stones embedded in their host rock – transformed the search for diamonds' origins. In 1866, children playing in a dry streambed of South Africa's vast central semidesert found the first tantalizing gem, a robin's-egg-sized blue–white pebble of more than 20 carats.9 Within a year of that find, the "shiny pebble" was known to the world as the Eureka. Two years later, the 83.5–carat Star of South Africa eclipsed Eureka and captured the imagination of fortune seekers around the world. By the end of 1870 an epidemic of diamond fever had brought more than 10 000 miners to places with names now embedded in mining history: Jagersfontein, Bultfontein, Dutoitspan, and, of course, Kimberley.

For a time diamond mining was a haphazard affair. At first it appeared that the South African deposits centered around river- and streambeds, similar to concentrations in India and Brazil. The first large diamonds came from surface workings, such as those along the Vaal River Valley northwest of Kimberley. But prospectors soon located concentrated deposits of diamonds in soft yellow rock that proved to be the weathered necks of ancient volcanoes. A new kind of diamond mine had been found, and new methods were imposed to exploit it.

Each miner at Kimberley was assigned one or two claims, each about 35 feet square (fig. 2). Each claim became a pit, sometimes as much as 100 feet deep. As some claimholders labored more diligently, or more recklessly, than others, the Kimberley topography grew more and more chaotic. Historic photographs of the early workings show a nightmarish landscape of sheer rock walls and deep holes, festooned with a complex webbing of ropes and pulleys, each belonging to a different claim. This unstable situation could not last long. Shallower workings collapsed onto deeper ones and the deeper pits began to flood, forcing a more communal effort. By the 1880s, visionary entrepreneurs like Cecil Rhodes, who began his fortune by selling water pumps to the miners, had transformed the thousands of individual diggings into a central-ized, mechanized South African diamond industry that surpassed the output of Brazilian and Indian diamond fields many times over.

Fig. 2 Kimberley Mine began as hundreds of separate claims. Different miners worked at different rates, producing a chaotic landscape of diggings. (Courtesy of F. R. Boyd.)

Millions of carats per year poured into a growing, middle-class, mostly American market, with the South Africans carefully controlling the supply and price of the precious stones.

Amid the diamond hunting frenzy at Kimberley, geologists eagerly scrutinized the distinctive matrix that held the precious crystals – a rock they called kimberlite.[10] They had never seen such a deposit, and soon they began to read the awesome testimony of the rocks. Kimberlite spoke of an origin deep within the earth, where intense heat and unimaginable pressures turn everyday stones to new, dense forms. The rock's geological situation, in large cone-shaped masses cutting across shattered country rock, attested to violent volcanic eruptions unlike anything recorded in human history.

Diamonds form when thousands of atmospheres of pressure combine with white-hot temperatures to coax carbon atoms into a dense atomic arrangement. Natural diamonds are rare because they can grow only deep within the earth – 100 miles or more down, where pressures reach hundreds of tons per square inch. But this discovery of diamonds' origins raised another question: If diamonds can only form at immense pressure – if soft, black graphite is the only stable variety of carbon at the earth's surface – how could diamond crystals have survived the epic journey from the earth's depths?

We now know that diamonds that take too long to ascend from the depths do in fact convert to graphite. Geologists have discovered incredibly rich deposits of what were once diamonds, but are now diamond-shaped masses of ordinary graphite.[11] These rock outcrops that once held up to fifteen percent of gems 10–carats and more – hundreds of millions of dollars of once-perfect diamonds in every cubic foot – confirm that many gems have been destroyed during their long upward journey.

The small fraction of diamonds that do reach the surface unaltered must experience a rapid release of pressure and temperature. The trip from below must be fast and violent, via volcanic eruptions of a fury and speed beyond our experience. A mass of partially molten diamond-bearing rock may sit deep within the earth for a billion years, but it can blast to the surface in less than an hour. Hot, upwelling magma weakens overlying rock, then molten rock mixes with water, carbon dioxide, and other pressurized gases, until microscopic cracks permeate the overlying rock.

Suddenly the rock gives way and the hot diamond-bearing mass erupts in a cataclysm beyond imagining. Magma from a hundred miles deep rises faster and faster, pushing through solid rock, rocketing to the surface. Hot expanding gases rush out with explosive force as the molten mass thrusts upward; the blast pulverizes much of the rock and its treasure of precious stones, scattering debris over thousands of square miles, leaving behind a cone-shaped plug of diamond-bearing rock that fills the cavity of the blast. Kimberlite, cradling its diamond hoard, cools in this near-surface environment. For millions of years wind and weather slowly, grain-by-grain, erode the rock away, and every so often a small, hard gemstone works its way free, washing into a nearby river or stream.

Kimberlite taught scientists what they had to do to make diamonds in the laboratory: squeeze carbon, and heat it unmercifully, to mimic conditions in the depths of the earth. But how could humans duplicate the pressure of a column of rock many miles high? How could they provide the temperature of a blowtorch to pressurized carbon atoms?

Those were tricks worth learning.

Notes

1. The largest known natural diamond crystal, the 3106-carat Cullinan, was 10 cm in maximum dimension and appeared to be a fragment of a much larger roughly octahedral crystal. Given the extremely rare and chance sampling of diamonds from the mantle, it is not unreasonable to assume the existence of diamonds up to a factor of 10 larger.

2. The exceptional physical properties of diamond are discussed in *Diamond* by Gordon Davies (Bristol: Adam Hilger, 1984). A useful review of natural diamond, with numerous references, is provided by D. N. Robinson, "The characteristics of natural diamonds and their interpretation." *Minerals Science and Engineering* **10**, 55–72 (1978).

3. A number of fine books review the history and lore of famous diamonds. These include, George Harlow (Editor), *The Nature of Diamonds* (Cambridge: Cambridge University Press, 1997); Ian Balfour, *Famous Diamonds* (London: Collins, 1987); Joan Y. Dickinson, *The Book of Diamonds* (New York: Avenel, 1965).

4. For more information on light–matter interactions and the optical properties of crystals refer to standard texts or encyclopedia articles on optics. See, for example, F. Donald Bloss, *An Introduction to the Methods of Optical Crystallography* (New York: Holt, Reinhart and Winston, 1967).

5. An overview of techniques for diamond cutting, faceting, and polishing are described by Joan Y. Dickinson in *The Book of Diamonds* (New York: Avenel, 1965).

6. Issac Newton, *Optiks* (London, 1704).

7. A.-L. Lavoisier, *Memoire Academie des Sciences* 1772, pp. 564, 591.

8. Smithson Tenant, "On the nature of diamond." *Philosophical Transactions of the Royal Society* **87**, 97 (1797). For a biography of Tenant, see Anonymous, "Smithson Tenant, F.R.S. (1761–1815)." *Royal Society of London Notes and Records* **17**, 77–94 (1962).

9. For a history of the discovery of the world's principal diamond sources, see Alfred A. Levinson, "Diamond sources and their discovery," in George Harlow (Editor), *The Nature of Diamond* (Cambridge: Cambridge University Press, 1997), and references therein. A useful history of South African diamond mining is provided by A. A. van Zyl, "De Beers' 100 – a special feature." *Geobulletin (of the Geological Society of South Africa)* **31**, 23–50 (1968). For an entertaining, though somewhat dated, popular account of the development of South Africa's diamond industry, see Emily Hahn's *Diamond* (New York: Doubleday, 1957).

10. Perhaps the first mention of the role of pressure in natural diamond formation appears in H. Carvill Lewis's, "On a diamantiferous peridotite, and the genesis of diamond." *Geological Magazine* **4**, 22–24 (1887). He states: "It seems evident that the diamond-bearing pipes are true volcanic necks . . . and that the diamonds are secondary minerals produced by the reaction of this lava, with heat and pressure, on the carbonaceous shales in contact with and enveloped by it."

11. D.G. Pearson, G.R. Davies, P.H. Nixon, and H.J. Milledge, "Graphitized diamonds from a peridotite massif in Morocco and implications for anomalous diamond occurrences." *Nature* **338**, 60–62 (1989).

CHAPTER 2

ATTEMPTS

> " Diamonds," he began . . . "No one yet has hit upon exactly the right flux in which to melt up the carbon, or exactly the right pressure for the best results. . . . Suppose one to have at last just hit the right trick, before the secret got out and diamonds become as common as coal, one might realize millions. Millions!"
>
> H. G. WELLS, "THE DIAMOND MAKER," 1894[1]

MANY MEN HAVE TRIED THEIR HAND AT diamond making: brilliant men, ambitious men, men driven by vision and men possessed by greed. The Frenchman C. Cagniard de la Tour claimed to have grown diamonds from solution in 1828, but his crystals were nothing more than aluminum and magnesium oxides – the fool's gold of diamond makers. In the same year, J. N. Gannal made a similar claim, but no one could reproduce his results.[2] In the 1850s, Charles Despretz announced the synthesis of diamond in an electric arc; others disproved his findings.[3] Within a century of Tennant's discovery that diamond is pure carbon, a dozen scientists had staked their claim in the diamond-making game. All of these early efforts focused on temperature alone to transform black carbon into diamond. Before Kimberley, the central role of pressure was unknown.

Of all the men who tried to make diamonds – the gifted researchers and ignorant fools, the seekers of truth and the outright frauds – none failed more spectacularly than Scottish chemist James Ballantyne Hannay.[4]

By the 1850s, chemistry had grown from its awkward infancy to an exuberant youth, buoyed by discoveries of magical wonder and extraordinary utility. New chemical elements, new compounds, new techniques for processing raw materials all added to the burgeoning repertoire of the chemist. New fuels, new pigments, new medicines, and thousands of other novel chemicals devised by curious scientists inexorably altered life styles in the self-confident material world.

The power of chemistry to transform matter had fascinated Hannay since his childhood in Glasgow, where as a precocious ten-year-old he had manufactured his own fireworks. Though largely self-trained in his modest home laboratory, Hannay acquired sufficient chemical expertise to be elected a Fellow in the Royal Society of Edinburgh in 1876, at the age of only twenty-one.

Hannay became the manager of a Glasgow chemical firm, which allowed him to continue his chemical inquiries on the side. Much of this exploration focused on the solvent properties of fluids at very high temperatures or pressures. He found that, in general, the higher the temperature or pressure of a gas or liquid, the greater its dissolving powers. In these experiments, Hannay practiced classic chemistry – mixing chemicals in solutions to see what happened, in the tradition of centuries of laboratory study. But, as so often happens in science, one seemingly straightforward experiment led Hannay in a completely unexpected direction. While attempting to dissolve sodium and lithium metals in common liquids like melted paraffin wax and oils, he found that the liquids sometimes broke down, releasing hydrogen gas and depositing the carbon as a hard, scaly coating that could scratch glass. The tough layer looked to Hannay like diamond.

This surprising result gave Hannay the idea for a new diamond synthesis strategy, and in the late 1870s, years before geologists understood the origins of Kimberley diamonds, he became obsessed with the challenge. His first synthesis attempts relied on the tried and true methods of the chemist's craft. He employed the most basic and rudimentary crystal-growing technique, dissolving the desired chemical – in this case carbon – in a liquid and waiting for the diamond to crystallize.

A variety of experiments employing simple liquids at normal pressures yielded only graphite. It was at this time that South African geologists had discovered the importance of pressure in the formation of natural diamonds. Hearing of their discovery, Hannay decided to try to mimic the earth's conditions. His heroic and foolhardy attempts involved filling lengths of iron tubing with carbon-rich oil and lithium, sealing both ends, and placing the cylinders in a massive six-foot-long furnace with foot-thick masonry walls. The tubes were fired for many hours at red heat (perhaps 900°C) – usually until they exploded from excess pressure.

High-pressure research was *terra incognita* to Hannay and his contemporaries, and his first abortive experiments were hampered by many technical difficulties. His first concern was sealing an iron tube so that it would withstand high internal pressures. Screw fittings, the most obvious choice, invariably leaked. He then tried inserting an iron ball into the end of the tubes and crimping the ends shut to form a seal that became progressively tighter as pressure increased. Unfortunately, the ball seal became a lethal high-velocity projectile at high temperature, when the crimped rifle barrel softened; "the iron yielded and the ball was driven out with a loud explosion."[5] Reluctantly Hannay resorted to difficult welded seals for each of his subsequent experiments. It was no mean feat to obtain a complete weld while preventing the highly volatile hydrocarbons from escaping. "It is only one man in a hundred who can perform the operation with invariable success," Hannay wrote – a pronouncement that reflects the lesson familiar to all experimental researchers, who understand that the most valuable laboratory asset is often the skilled technician.

No diamonds were found in experiments using ordinary one-inch-diameter tubes with half-inch bores, so Hannay began using thicker and thicker cylinders. A few experiments were tried with two-inch barrels, which burst with regularity, so tubes close to three inches in diameter were substituted. They too exploded or leaked. Finally he resorted to four-inch-diameter tubes, which blew up at even higher pressures, with even more violence.

In a typical experiment, Hannay wrote, the tube "exploded with a great noise, and knocked down the back and one of the ends of the furnace, leaving the whole structure a wreck." In the rare experiment that didn't leak or explode, the risk to Hannay and his workers was perhaps even greater. The highly pressurized welded cylinder had to be drilled open by hand, the technician was exposed to considerable danger when pressurized gas finally escaped with a loud report and violent rush. Hannay expressed relief that one of his workers narrowly escaped injury when a tube ruptured and shattered in a spray of shrapnel as it was being drilled. "The continued strain on the nerves," he noted, "watching the temperature of the furnace, and in a state of tension in case of an explosion, induces a nervous state which is extremely weakening, and when the explosion occurs it sometimes shakes one so severely that sickness supervenes."

During one short period, Hannay lamented, "Eight tubes failed through bursting and leaking, and one of the explosions . . . destroyed a part of the furnace and injured one of my workmen." Of eighty laborious high-pressure experiments, seventy-seven were failures. In most cases run products were lost through leakage or were scattered about the furnace in an explosion. For many months the few runs that weren't dispersed produced only graphite if anything at all. But in early 1880, Hannay finally, joyfully, found traces of diamond. Tiny clear crystals, too hard to grind in a mortar and impervious to acid, were recovered from solid residues in three tubes. Chemical tests convinced Hannay that he had made diamonds. The weary researcher celebrated, wrote up and published details of his success in the Royal Society's *Proceedings* for 1880, and went on to other things.

At the time most observers assumed that the diamond synthesis problem had been solved, albeit in a noncommercial way. Hannay's triumph was accepted as just another in the flood of chemistry breakthroughs of the late nineteenth century. But we now know that something was wrong with his results. Recent analyses of Hannay's "synthetic" diamonds, which are preserved in the collections of the British Museum, prove without question that they are natural, not synthetic.[6] These tiny fragments of history, the only surviving "synthetic diamonds" from the many claimed before World War II, contain a telltale impurity. Natural diamonds, held deep in the earth for millions of years, often develop thin plate-like concentrations of nitrogen atoms dispersed through the carbon crystal lattice. No rapidly formed synthetic diamond could display this gradual growth feature, nor is it likely that such orderly impurities would form over any human time scale.

There can be no doubt that Hannay's diamonds came from a natural source. Some historians have denounced Hannay as a fraud, though others of more charitable disposition suggest that Hannay's workmen, fearing for their employer's safety as well as their own, planted the natural diamond bits to end the hazardous experiments. Chances are we will never know for sure.

* * *

Hannay's experience with exploding iron tubes and shattered furnaces was hardly unique in the search for synthetic diamonds. Many nineteenth-century researchers took grave risks in their pursuit of knowledge. Scientists are seldom portrayed as heroes who risk life and limb for a higher cause, but chemistry in the nineteenth century held many hazards. Ferdinand-Frederick-Henri Moissan, Nobel Prize winning chemist and would-be diamond maker,[7] almost died in his crusade to become the first to isolate the lethal gas fluorine.

Moissan and other chemists of the nineteenth century realized that just a few dozen different building blocks – the chemical elements – form our world, and they placed the highest priority on cataloging and isolating those elements. Some naturally occurring "native" elements, like gold, silver, copper, and lead, had been known for thousands of years, while others, like oxygen, aluminum, magnesium, and hydrogen, were easily derived from common compounds as soon as a few simple chemical tricks were learned. Dozens of other kinds of atoms remained hidden and unsuspected as trace elements in rare minerals. Fluorine was a maddening exception. Common "fluoride" minerals of calcium and sodium were loaded with the distinctive light element, but fluorine atoms were so tightly bonded to the others that it seemed impossible to separate and purify the stuff. It was as if the chemists had been invited into a fabulous castle, each glorious room containing a new precious element, except that one of the grandest chambers – the room containing fluorine and its treasures – remained locked and taunting, and booby trapped, to boot. Many tried to unlock the secret of fluorine and paid a terrible price.[8]

When Moissan began his most famous research in 1884, toxic fluorine and its compounds had already killed or injured a half-dozen respected scientists. Early in the century the famed English chemist and lecturer Sir Humphry Davy and the rival French team of Joseph Louis Gay-Lussac and Louis Jacques Thenard both tried and failed, only to suffer debilitating eye irritation and respiratory inflammation from their fruitless fluorine work. The Irish brothers George and Thomas Knox also suffered horribly when they were exposed to hydrofluoric acid. Thomas almost died, while George spent three years recovering his health in Naples. George Gore of London managed to isolate a small quantity of the gas, but it almost immediately combined explosively with hydrogen, destroying part of his laboratory. And they were the lucky ones. Both Jerome Nickles of France and Paulin Louyet of Belgium, though fully aware of the dangers, died horribly in their laboratories – asphyxiated as their burned and blistered lungs were destroyed by the corrosive gas.

It was a dangerous game these chemists played, but Moissan accepted the risks. In a crude, poorly ventilated laboratory in Paris he passed strong electrical currents through a mixture of hydrofluoric acid and fluorine salts in a costly apparatus constructed largely of platinum. He, too, suffered from fluorine poisoning, and shortly before his death at the age of fifty-four he blamed these chemical investigations for shortening his life by a decade. But in 1886, after almost three years of labor, he succeeded in purifying the nasty, corrosive greenish-yellow gas. Then, with his unique and steady supply of the element, isolated by the process still used in industry today, he was able to create a variety of totally new chemical compounds spawned from fluorine's unrivaled reactivity.

This work greatly enhanced Moissan's reputation, and led to his appointment to a prestigious academic position (in toxicology, no less). But it also turned out to be an excellent foundation for his next research problem – the synthesis of diamond.

Moissan began his painstaking, and considerably safer, diamond work with the meticulous care of a craftsman. To make diamond in the laboratory, he reasoned, one must know as much as possible about how diamonds arise in nature. To this end, he gathered three very different samples of diamond-bearing material. He examined kimberlite from South Africa in the greatest detail, finding large numbers of previously unsuspected microscopic diamonds embedded in it. He dissolved almost five tons of Brazilian diamond-bearing gravel in strong acids to concentrate about two carats of fine diamond particles. Finally, he extracted two tiny diamonds from a piece of the Canyon Diablo iron meteorite, found near the great meteor crater in Arizona. In all these samples, iron was diamond's common companion. Moissan thought a metal like iron was the key to diamond synthesis.

To find out, Moissan designed an extraordinary new oven capable of melting iron and most other metals.[9] His electric arc furnace achieved record temperatures of 3000°C (about 5000°F) by passing a powerful electrical current through graphite, which glowed white with heat like the metal strips in a space heater.

The most famous photograph of scientist Moissan reveals much about the man (fig. 3). Taken early in the twentieth century, the image captures an elegant, keen-eyed academic, formally attired and studiously erect, yet standing in a crude brick-walled basement room next to a rugged laboratory workbench. Though his chief claim to fame and the basis for his Nobel Prize was the separation of fluorine, Moissan was intensely proud of his furnace. The crude pile of fire bricks and thick electric power cables dominate the central third of the photograph; its

Fig. 3 Henri Moissan stands beside his electric arc furnace, in which he believed he had synthesized diamonds (R. M. Hazen).

inventor stands to the left, one arm in familiar contact with his imposing device.

Moissan's thick full beard and grand moustache denote an honored
professor's self-confidence, and perhaps an avowed experimentalist's
pragmatism as well. The synthesis of crystals is as much an art as a
science. You can perform the same crystal-growing experiments
dozens of times; sometimes crystals form, sometimes they don't. The
experimental chemist knows that one key to successful synthesis is
often just having a tiny bit of the right impurity to provide a starting
place – a "nucleation site" for crystal growth. But how do you sprinkle
your experiment with such microscopic catalysts? Experimental chemists learned long ago that nothing works better than a full beard for collecting and redistributing a laboratory's chemical dust. A beard like
Moissan's, exposed to a long career in chemical laboratories, must have
been full of exotic stuff. Maybe it's only a chemist's superstition, but
Moissan did create a lot of new crystals.

In his first set of diamond-making experiments Moissan used his
furnace to melt iron and other metals in a crucible, dissolved carbon (in
the form of burned sugar) in the liquid metal, and then let the carbonized mixture cool (fig. 4). The procedure must have been spectacular,
for at the extreme temperatures of the electric arc furnace graphite
glows intensely white and iron erupts in a shower of sparks. After the

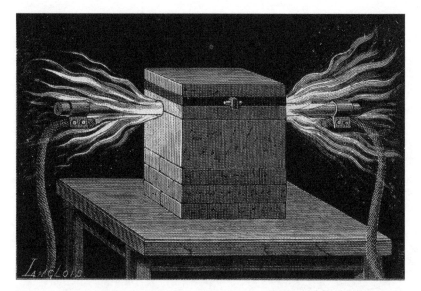

Fig. 4 At full power, Henri Moissan's electric arc furnace, pictured in this illustration from his 1904 monograph, *The Electric Furnace*, produced a spectacular display of sparks and flame as it reached temperatures of 3000°C. Moissan believed that he had synthesized diamonds in his apparatus (R. M. Hazen).

blasted sample cooled, acids were used to dissolve the metal away, leaving behind a residue of ordinary black carbon – graphite – in every experiment.

Moissan realized that pressure represented a key missing ingredient, but he lacked the equipment to subject samples to simultaneous high temperature and pressure. His ingenious solution was to let the cooling, contracting metal provide the pressure itself. Carbonized liquid iron was prepared as in the earlier trials, but this time it was rapidly chilled in cold water, which caused the outer portions of the iron to solidify almost instantly. The sudden contraction of this cooling iron jacket, it was hoped, would generate the necessary pressure to induce diamond growth. To increase the cooling rate even more, Moissan dropped the iron into "baths" of iron filings or even molten lead – a substance hundreds of degrees cooler than the iron, and one that actually caused the iron's temperature to drop more quickly than in water.

Many of these primitive high-pressure experiments produced microscopic carbon particles (fig. 5), never larger than a few hundredths of an

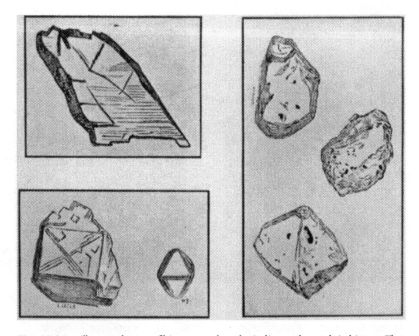

Fig. 5 Moissan illustrated many of his supposed synthetic diamond crystals in his text, *The Electric Furnace* (R. M. Hazen).

inch, that appeared to have the distinctive properties of a diamond. The particles displayed high density, scratched other hard materials, and some even had the rough octahedral crystal form of natural diamond. Most important, Moissan's diamonds were said to burn, leaving a residue of carbon dioxide.

Henri Moissan went to his death in 1907 believing that he had made diamonds, and few challenged the distinguished chemist's assertion while he lived. But there were doubts, and within a few years, after several respected scientists failed to reproduce his claims, Moissan's experiments were all but dismissed. Many years later, Moissan's widow expressed her belief that one of the professor's devoted assistants may have "sprinkled" the runs with a few natural diamond fragments "to please the old man."

Modern recreation of Moissan's procedures reveal a different explanation for the diamond-like products. Each synthesis run yielded several tiny hard crystals, very much like some of the ones recovered by the Frenchman. Those crystals have been identified as silicon carbide

– a shiny, abrasive known commercially as carborundum. Natural silicon carbide has subsequently been given the name moissanite, to honor the man who first inadvertently made it, instead of diamond.

* * *

Moissan firmly believed that his experiment had succeeded, and he meticulously spelled out the diamond-making recipe in his monograph *Le Four Électrique*, translated into English as *The Electric Furnace* in 1904. In that same year another Frenchman, Monsieur Lemoine, capitalized on Moissan's publicity and claimed to have made large diamond crystals in his own modified electric furnace. A diamond expert from Kimberley was dispatched to Paris, where he watched Lemoine "transform" ordinary charcoal into a diamond. Unfortunately, Lemoine's supposedly man-made stone had all of the unmistakable surface markings of a river-washed gemstone. The unscrupulous Lemoine was subsequently arrested and tried for fraud.

But not all of Moissan's followers were frauds, however. His widely acclaimed synthesis experiments served as the starting point for the work of many other respected scientists. One of the most ardent and unquestioning believers was Professor J. Willard Hershey, who conducted more than a decade of misguided diamond-making experiments (fig. 6) as classroom exercises for his chemistry students at McPherson College in Kansas.[10] Hershey even dedicated his own *Book of Diamonds* to "all my students who have had any part in helping to make synthetic diamonds under my instruction."

Also among the avid readers of Moissan's *The Electric Furnace* was Sir William Crookes (fig. 7), a towering figure in nineteenth-century British science, remembered as much for his baffling advocacy of the occult as for his solid technical contributions.[11] Crookes's impressive catalog of accomplishments includes discovery of the element thallium and invention of the radiometer. He also designed an improved vacuum pump, discovered new applications for spectroscopy, and made important advances in photography and in methods for extracting valuable metals from ore. He founded the influential journal *Chemical News* and was its sole editor for almost half a century. He served as president of the Royal Society and many other organizations and enjoyed a knighthood and numerous other public and professional honors. Yet his obsession with spiritual phenomena, his endorsement

Fig. 6 Charles Wagner, an assistant to Professor Williard Hershey, conducts diamond synthesis experiments at McPherson College in Kansas. Hershey and his colleagues employed
the electric arc technique of Moissan (R. M. Hazen).

of certain well-known mediums, and his belief in a psychic force that
could modify gravity and cause other physical manifestations, compli-
cate the portrait of a man who achieved so much in the "rational" world
of science.

In a career so sweeping and vibrant, Crookes's few diamond-making
experiments appear to rate little more than a footnote, but his novel
strategy foreshadowed the twentieth-century diamond-making industry. In 1896, in his sixty-fourth year, Crookes spent nearly a month studying the mineralogy and geology of Kimberley.[12] He saw diamond
mines at first hand and investigated the natural chemical origins of the
gems, grasping the key role of pressure in diamond synthesis. Then for
almost a decade other research, particularly in the youthful field of
radioactivity, distracted him, but the 1904 publication of Moissan's
treatise and Crookes's second South African trip in 1905 galvanized
him to try his hand. His attempts to duplicate Moissan's contracting-
iron method seemed to work well, for Crookes reported "veritable diamonds" (probably silicon carbide), with all the distinctive attributes of
the natural stone.

Fig. 7 William Crookes (R. M. Hazen).

Part of the challenge of chemistry, however, is to find new paths to synthesis. Moissan's procedures were old news, and noncommercial to boot. Crookes knew that a cheap method of diamond synthesis would have great commercial applications, so rather than dwell on what worked, he elected to try a new approach, taking Hannay's sealed-tube method a step further. The highest temperatures and pressures available in Crookes' day had been obtained by Sir Andrew Nobel, who studied the effects of explosives. Noble had packed steel tubes with gunpowder or cordite and heated them until they ignited with tremendous explosive force, briefly attaining pressures of 8000 atmospheres at more than 5000°C.[13] Crookes was given the chance to study the solid residues from these blasted tubes. He wrote, "After weeks of patient toil I removed the amorphous carbon, the graphite, the silicon, and other constituents." Subsequent treatment by heat, nitric acid, and sulfuric acid yielded a residue of tiny transparent crystals. "Chemists will agree with me that diamonds only could stand such an ordeal; on submitting them to skilled crystallographic authorities my opinion is confirmed.... In these closed vessel experiments we have another method of producing the diamond artificially."

For a second time Crookes believed he had produced diamond, although modern researchers are convinced otherwise. None of his synthetic material has been preserved, but we know that the explosive method could not possibly have produced sufficient temperatures and pressures. Chances are that silicon carbide was, once again, the imposter.

* * *

Most of the would-be diamond makers of a century ago dedicated their energies to one novel idea or another for a few years and then went on to tackle a completely different problem. Sir Charles Algernon Parsons (fig 8), son of an Irish Earl and privileged recipient of the best education money could buy, was different.[14] Diamond making became his lifelong obsession, and he tried every scheme he could think of to get the job done.

Persistence was certainly one of Parsons's strong suits. Though shy and introspective by nature, he could be doggedly determined in his efforts to publicize and promote his work. His chief claim to fame was the 1884 invention of a steam turbine for ship propulsion. This new power source was both efficient and reliable, but its implementation

THE HON. SIR CHARLES PARSONS, O.M., K.C.B., M.A., D.Sc., F.R.S.

Fig. 8 Sir Charles Parsons (R. M. Hazen).

was blocked for more than a decade by legal technicalities and a tradition-minded British Royal Navy. Parsons turned the tide in 1897 when he rocketed his own ship, the *Turbina*, at an incredible thirty-plus knots through the Diamond Jubilee fleet review while Queen Victoria and her wide-eyed admirals looked on. The dramatics paid off, for within a few years Parsons's steam turbines became the norm for most new British ships, including the mighty *HMS Dreadnought* and Cunard's magnificent *Lusitania*.

Experiments on steam turbines provided Parsons with firsthand experience in controlling high temperatures and pressures, and he applied that experience to diamond making as early as the mid-1880s. In his first series of experiments on pressurized carbon he subjected either wood charcoal or coke, a carbon-rich by-product of baked coal, to extreme electrical currents approaching 100 000 amperes, and pressures to about 4000 times atmospheric – a force of about thirty tons per square inch.[15] The yellowish crystal mass produced in these experiments suggested to Parsons the characteristics of diamond and inspired a long succession of similar attempts that focused on passing high electric currents through graphite.

Parsons's diamond-making experiments, though never successful, captured the public's imagination and quickly entered American popular culture. Howard Garis, alias Victor Appleton, based *Tom Swift Among the Diamond Makers*, the seventh volume of his immensely popular Tom Swift series of boys' adventure tales, on Parsons's ideas.[16] Published in about 1911, just a few years after Parsons described results of hundreds of attempts to synthesize diamonds using electric current, the book invokes a clever modification of the electric arc method in which lightning is used to transform a carbon mixture into precious gems.

In the novel, Tom Swift and his companions help a distraught inventor whose secret diamond-making process has been stolen. They sneak into a hidden mountain cave and watch the illicit preparation of a mixture of carbon and "other substances," ground and molded into small spheres, which are baked in an oven, placed into a steel box, and sealed tight before massive electric cables are attached to two electrodes on the box. The mountain, it turns out, is so rich with iron ore that it serves as a natural lightning rod. The diamond makers need only wait for the next thunderstorm for a diamond-forming jolt of energy. Unfortunately, the same awesome energy that fuses the diamonds also destroys the mountain cave in a dramatic explosion and sheets of blue flame in the book's improbable climax. Tom Swift and his companions barely manage to escape with their lives, but not before they scoop up a handful of synthetic gemstones, one of which Tom presents to his delighted girlfriend, Mary Nestor, in the book's closing scene.

Appleton's tale of adventure may have become the most enduring consequence of Parsons' electric experiments, but after exhausting that procedure he was by no means through. He attempted to modify Moissan's chilled iron method by compressing a molten mass of carbon-saturated iron to 12 000 atmospheres. He tried melting graphite at extreme pressures and temperatures, in the hopes it would cool as diamond. He mixed carbon with other elements – even with kimberlite rock. He even fired high-velocity bullets into carbon-rich material to make the gems by impact.

Charles Parsons was a careful and conscientious scientist who dutifully recorded the conditions and results of thousands upon thousands of diamond-making experiments, spanning thirty years of effort. He methodically isolated suspected diamond crystals and preserved them on glass slides for others to examine. For most of his career he was thoroughly convinced that he was making diamonds. But, sadly, as chemists analyzed and rejected sample after sample as merely carbides or simple oxides of aluminum and magnesium – all logical by-products of his experimental methods – Parsons came to realize that his dream had eluded him. After decades of labor all the results were negative, and Parsons eventually admitted to friends and colleagues that he, and all others, had failed. At the time of his death in 1931, not one reproducible experiment by any of a century of wishful diamond makers had succeeded.

Hannay, Moissan, Crookes, and Parsons, along with many contemporaries, all evidently failed to make diamonds, but their research was neither flawed nor useless. All scientific progress is built on the hard-won knowledge of previous generations, and these scientists had taken the first steps in striving to achieve the pressures and temperatures that would eventually accomplish this elusive goal. They also proved beyond a doubt that making diamonds would not be easy, and with each failure by another distinguished researcher, the stakes in the diamond-making game rose higher.

* * *

While Hannay and others engaged in their futile attempts to synthesize diamond, other scientists were learning more about what made the substance unique. Albert Einstein relied on diamond data to support his 1907 theory of specific heat, the energy associated with atomic vibrations.[17] Every crystal stores two kinds of energy. First, there is the bond energy itself – the potential energy of atoms held together by the attraction between negatively charged electrons and positively charged atomic nuclei. Everything you touch – the air you breathe, the food you eat, the clothes you wear – exists because of this stored bonding energy. The second kind of energy stored in the crystal is the energy of atomic motion, or heat. Every atom is in constant motion, and in a physical world motion means energy – kinetic energy.

These two kinds of crystal energy – binding and heat energy – obey different kinds of rules. Binding energy changes only slightly with temperature: the carbon-to-carbon bond energy of diamond at freezing temperatures or at red-hot heat is almost exactly the same. Heat energy, on the other hand, is directly dependent on temperature: hot matter has a lot more heat energy than cold. Einstein and his contemporaries worried about the exact nature of the heat–temperature relationship, and that's where diamond entered the picture.

Some scientists thought that an atom's energy can assume any value. If this is true, and energy varies smoothly and continuously, as in the classic theory of heat, then the graph of heat energy versus temperature has to have one distinctive curved shape. If, on the other hand, we live in a world where energy comes only in tiny discrete bundles – quanta – then the graph of heat versus temperature will show a rather different curve at low temperatures near absolute zero (about $-273°C$). Einstein assumed the quantum model and predicted the distinctive shape this second curve would have. Diamond, because of its high heat conductivity, was the perfect test case: it holds so little heat energy that quantum effects persist at relatively high temperatures. Experiments showed that diamond matched Einstein's predicted behavior almost perfectly, thus buttressing the revolutionary quantum theory.

* * *

Energy is just one of many attributes of our physical world that is quantized; matter, too, comes in discrete bundles called atoms. Before the discovery of x-rays, no one had seen direct evidence for these physical building blocks. No one knew their exact size or how they filled space to make reality. But with the invention of x-ray crystallography in 1912, chemistry entered the modern era. X-rays, the same radiation that probes your bones and tissues at the doctor's office, enabled scientists to determine the very architecture of our material world.

The principle of x-ray crystallography is simple enough. Every crystal features layer upon layer of atoms, repeating almost endlessly. X-rays striking a crystal will scatter off these layers, much as a beam of light will reflect in different directions off of an irregular surface. But if the x-ray beam is directed at a certain angle, and the atomic planes are carefully aligned, a sharp stream of x-rays – diffracted x-rays – flies from the crystal. By measuring the exact directions and intensities of these diffracted beams, crystallographers can calculate the size and arrangement of the atoms in the crystal.

In 1913 the father and son team of William and Lawrence Bragg employed these newly discovered principles of x-ray diffraction to deduce the elegant atomic structure of a diamond.[18] They found an arrangement in which every carbon atom is surrounded by a pyramid of four others – a dense interlocking network that contrasts sharply with the layered arrangement of graphite, in which each carbon has only three neighbors closely arrayed in a plane (fig. 9). Little wonder that graphite and diamond behave so differently. The carbon-to-carbon bonds within the layers of the graphite structure are quite strong, so the atomic sheets are tough and resilient, but the bonds between atom layers is much weaker. Other atoms can exploit this weakness; they enter the interlayer regions and allow the layers to slide across each other to provide an excellent, soft lubricant. Diamond, by contrast, has a rigid, three-dimensional structure, like cross-braced steel girders of a trestle bridge. Carbon atoms are densely packed in a diamond, which has more atoms in a given volume than any other solid.

Scientists wondered what extreme temperatures and pressures might be required to form dense diamond instead of graphite. The science of thermodynamics, which helps chemists relate a material's stability to the energy of its atomic arrangement, provided the answer. Chemists learned that atoms always try to arrange themselves in a way that minimizes their stored energy. Think of different chemical substances as boulders strewn around a mountainous valley. Some rocks sit precariously on the mountain slopes, ready to tumble down at the slightest jostle. These stones have excess energy, called gravitational potential energy because it is so readily available and poised to be unleashed. Explosive chemicals behave in this same way – a slight jiggle

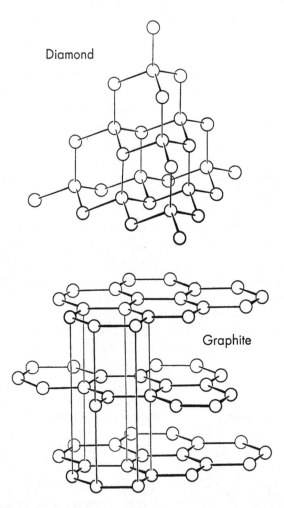

Fig. 9 The contrasting properties of graphite and diamond originate from their very different atomic structures. In diamond (top), carbon atoms are each surrounded by a pyramid of four other carbons, creating a three-dimensional atomic network of unparalleled strength. In graphite, on the other hand, each carbon is strongly bonded to three neighbors in layers, while bonds between layers are relatively weak.

and the atoms can rearrange into a more stable arrangement, releasing lots of energy (an explosion) in the process. Other boulders, however, rest firmly on the valley floor and nothing much will move them. These stable boulders, like everyday stable chemicals, have lower potential energy and, therefore, do not change spontaneously.

At any given combination of temperature and pressure, the structure of a substance most likely to form will be the one with lowest energy. Very high pressure, as you might expect, favors densely packed arrangements of atoms, like diamond; very high temperature tends to stabilize more open structures, like graphite. It's not always quite that simple, for other factors like the size of the atoms or the way they vibrate also play a role. But the basic idea holds true.

The science of thermodynamics took the guesswork out of predicting stable chemicals. If you know a few basic facts about the forms – their density, their atomic structure, and the energy tied up in their atomic bonds – you can often calculate which one will be most stable under a given set of conditions. Fortunately, the properties of diamond and graphite, the two competing forms of carbon, had been measured with great precision.

At very low temperatures and pressures the measurements were easy: scientists compared the amount of energy stored in graphite versus diamond by determining the heat released as a given weight of each substance burned. These experiments showed that the carbon–carbon bonds in diamond hold more energy than in graphite. Like a boulder poised on the side of a steep valley, diamond at room conditions seemed ready to tumble down to the more stable graphite form. At low pressure graphite is always the stable form.

The only reason diamond doesn't spontaneously convert to graphite is that carbon–carbon bonds are too strong to break without a large energy jolt. You can turn a diamond into graphite by burning it with a blowtorch, but that takes a lot of energy. It's as if a large boulder, though high above the valley floor, is sitting in a deep pit. You'd have to lift the boulder up over the lip of the pit before it could tumble down the slope to a more stable resting place.

At extremely high pressure, where atoms are forced together and atomic bonds are compressed, the story is quite different. The graphite structure collapses as weakly bonded layers of carbon atoms are forced uncomfortably close together. The internal energy of the squashed structure increases dramatically under pressure. But diamond, with its dense atomic arrangement, stabilizes at high pressure; its bonds hardly change at all. At high enough pressure diamond has the lower energy, and therefore is the more stable variety.

Thermodynamics allowed scientists to calculate whether diamond or graphite should appear for any combination of temperature and pressure.[19] During the period between the two world wars many scientists attempted to calculate the relative stabilities of graphite and diamond,

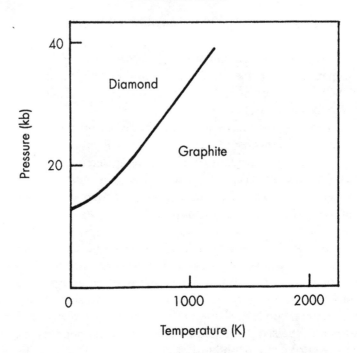

Fig. 10 The phase diagram of carbon shows whether diamond or graphite is the more stable form for any combination of temperature and pressure. In this early phase diagram, calculated by Frederick D. Rossini and Ralph S. Jessup in 1938, a line separates the high-pressure region where diamond is stable from the low-pressure graphite region. This figure guided the efforts of many would-be diamond makers.

and all agreed that at room temperature, up to a pressure of at least 10 000 atmospheres, graphite remains the stable form.[19] At the higher temperatures needed to make carbon atoms sufficiently mobile to change structure, pressures must be much higher. These researchers produced a carbon phase diagram – a graph of temperature versus pressure (fig. 10) that displayed which conditions would produce which form. The diamond field revealed the high temperatures and pressures at which diamond's energy is lower; the graphite field corresponded to lower energy for that mineral.

The message was unambiguous. If diamond was to be made in a laboratory by high-pressure methods, a lot more pressure was needed. And when it came to more pressure, Percy Bridgman was the man for the job.

Notes

1. H. G. Wells's short story, "The Diamond Maker," first appeared in 1894. It has subsequently been reprinted in many anthologies of his works.

2. J. W. Mellor, *A Comprehensive Treatise on Inorganic and Theoretical Chemistry.* (London: Longmans, Green, and Co., 1924). Chapter 39 (pp.710–771) treats carbon; Section 5 on "The synthesis and genesis of diamonds" (pp.730–738) provides a comprehensive summary of numerous diamond-synthesis attempts, with an extensive list of pre-1923 references.

3. Charles Depretz reported on his synthesis experiments in a series of articles that include: 'Fusion et volatilisation des corps refractaires. – Note sur quelques experiences faites avec le triple concours de la pile voltaique, du soleil et di chalumeau.' *Comptes Rendus* 28, 755–757 (1849); "Deuxieme note sur la fusion et la volatilisation des corps." Ibid. 29, 48–51 (1849); "Troisieme note sur la fusion et la volatilisation des corps." Ibid. 29, 545–548 (1849); "Quatrieme note sur la fusion et la volatilisation des corps." Ibid. 29, 709–724 (1849); "Cinquieme communication sur la pile. Quelques nouvelles experiences sur le charbon. Longueurs de l'arc voltaique." Ibid. 30, 367–376 (1850); "Addition a la note sur le charbon, lue dans la seance du 5 Septembre 1853." Ibid. 37, 443–450 (1853).

4. James B. Hannay's research on diamond synthesis was described in "On the artificial formation of the diamond." *Proceedings Royal Society London*, 450–461 (1880), and "Artificial diamond," *Nature*, 255–257 (1880).

5. Ibid., *Proceedings Royal Society London*, 453–457 (1880).

6. K. Lonsdale, "Further comments on attempts by H. Moissan, J. B. Hannay and Sir Charles Parsons to make diamonds in the laboratory." *Nature* 196, 104–106 (1962). See also: A. T. Collins, *Industrial Diamond Review*, December 1975, p. 434.

7. In addition to the excellent biographical essay in Dictionary of Scientific Biography, see Moissan's obituary in *Nature* 75, 419–420 (1907).

8. A contemporary history of fluorine research is found in A. E. Tutton's, "Isolation of fluorine." *Nature*, 179–183 (1887). See also pp.50–52 of J. Newton Friend's, *Man and the Chemical Elements* (New York: Scribners, 1953), and pp. 454–468 of Mary Elvira Weeks, *Discovery of the Elements* (New York: Journal of Chemical Education, 1948).

9. Henri Moissan, *The Electric Furnace* (translated by A. T. de Mouilpied). (London: Edward Arnold, 1904). Part III, "Artificial production of the diamond" (pp. 77–141), contains a detailed description of Moissan's experiments. Additional analysis is provided by: C.H.D., "The problem of artificial production of diamonds," *Nature* 121, 799–800 (1928).

10. J. Willard Hershey, *The Book of Diamonds: Their Curious Lore, Properties, Tests and Synthetic Manufacture.* (New York: Hearthside Press, 1940). Chapter 14, "How to make synthetic diamonds" (pp. 123–139), is a fascinating illustrated guide to the procedures used by Hershey and his students at the Department of Chemistry at McPherson College, McPherson, Kansas.

11. Biographical details of William Crookes's life are provided by E. E. Fournier d'Albe, *The Life of Sir William Crookes O.M., F.R.S.* (London 1923). See also P. Zeeman, "Scientific worthies: Sir William Crookes, F.R.S." *Nature* 77, 1–3 (1907), and the biographical essay by W.H. Brock in the *Dictionary of Scientific Biography*.

12. Sir William Crookes, *Diamonds*. (London and New York: Harper and Brothers, 1909). Crookes reviews the synthesis of diamonds in Chapter 9, "Genesis of the diamond" (pp.115–126), in which he credits Moissan for the first artificial production. Much of the volume describes Crookes's two trips to the Kimberley Mine in South Africa.

13. Sir Andrew Nobel, "Research on explosives. Part III. – Supplementary note." *Proceedings of the Royal Society London* **76a**, 512–514 (1905).

14. Rollo Appleyard, *Charles Parsons: His Life and Work*. (London: Constable and Co., 1933). See especially Chapter 9 (pp. 221–238), "Diamonds and bore-holes." Additional insights are provided by a collection of memorial letters: "The Hon. Sir Charles Algernon Parsons, O.M., K.C.B., F.R.S." *Nature* **127**, 314–316 (1931). See also: Fourth Baron Rayleigh, "Laboratory synthesis of diamond." *Nature* **151**, 394 (1943).

15. C. A. Parsons, "Some notes on carbon at high temperatures and pressures." *Proceedings of the Royal Society London* **79a**, 532–535 (1907).

16. Victor Appleton, *Tom Swift Among the Diamond Makers*. (New York: Grosset and Dunlap, *ca*.1911). The historical context of the Tom Swift adventures is provided by Bruce Watson's, "Tom Swift, Nancy Drew and pals all had the same dad." *Smithsonian Magazine* **22**, 50–61 (1991).

17. Albert Einstein, "Die Plancksche Theorie der Strahlung und die Theorie der spezifischen Warme." *Annalen der Physik, Leipzig* **22**, 180–190 (1907).

18. W. H. Bragg and W. L. Bragg, "The structure of the diamond." *Proceedings of the Royal Society London* **A89**, 277–291 (1913).

19. F. D. Rossini and R. S. Jessup, "Heat and free energy of formation of carbon dioxide, and of the transition between graphite and diamond." *Journal of Research of the National Bureau of Standards (USA)* **21**, 491–513 (1938). Other early estimates of the graphite–diamond phase boundary include: F. Simon, *Handbuch der Physik* **10**, 350 (1926), and O. I. Liepunskii, *Upspekh' Khim ii* **8**, 1519 (1939).

CHAPTER 3

THE LEGACY OF PERCY BRIDGMAN

" If intensely hot carbon – heated with an electric arc let us say – were subjected to an enormous pressure might not diamonds be formed? . . .
Would you send me a diamond so made?"[1]

MISS MARTIN TO PERCY BRIDGMAN, June 10, 1923

" I have thought a good deal about the possibility of making diamonds by
subjecting carbon to high pressures, and whenever I have had a new kind
of apparatus made that presented features that I had not tried before, I
have always surreptitiously put into it a little piece of carbon before applying the pressure. . . . I will certainly send you a diamond when I find how to
make them – the second one – the first has been promised for a long time
to Mrs. Bridgman."[1]

PERCY BRIDGMAN TO MISS MARTIN, July 4, 1923

ARMED WITH A VISION OF DIAMOND'S violent origins and a
growing body of research testifying to the methods that wouldn't
work, scientists were ready to tackle the synthesis problem with new
intensity. By the first decades of this century virtually everyone in the
diamond-making game agreed that pressure was the key to diamond
synthesis.

No one was more fascinated by pressure's transforming power than
Percy W. Bridgman, the Harvard physicist who ushered in the modern
era of high-pressure research by squeezing just about everything he
could get his hands on. He devised new ways to compress solids,
liquids, and gases between steel vise jaws, and he used novel techniques
to measure the properties of the compressed matter. Along the way, he
produced some amazing phenomena. Bridgman found new kinds of
ice by compressing ordinary water. He watched everyday gases like
nitrogen and carbon dioxide turn to crystalline solids under pressure.
He caused simple salts to transform into metals, and observed minerals
crushed to half their normal volumes. In a half-century of Nobel Prize-
winning research Percy Bridgman studied almost a thousand different

substances,[2] yet no single problem occupied more of his career or posed more of a challenge than his attempts to change graphite into diamond.

What drives a scientist to study matter at high pressure? For some it's the massive steel tools of the trade – machines that harness thousands of tons of force and can crush rock to dust or flatten an automobile into a few cubic feet of scrap metal. Surely some scientists are captivated by pressure's awesome power.

But there is something more subtle in the allure of squeezing matter. In a scientific world that focuses ever more intently on the incomprehensibly small and unimaginably brief phenomena of the quantum world, in a century in which relativity has turned our perception of time, space, and physical reality on its head, it is comforting to do science on real stuff that you can see and hold in your hand. There is a concrete satisfaction in taking a chunk of matter, placing it in the jaws of a powerful vise, and turning the screw.

Percy Bridgman understood that allure. As an American boy growing up in the 1890s, he accepted the country's prevailing view that inventors Thomas Edison and Alexander Graham Bell were among the greatest living physicists, and he embraced those inventors' empirical philosophy. As twentieth-century physics gradually became more abstract, Bridgman retained the belief that models of the universe had no meaning if they could not be tested by direct observation. The only meaningful questions, he said, were those that could be answered by running real experiments in a real world. And that is what he did.[3]

*   *   *

Percy Williams Bridgman was born in 1882 in Cambridge, Massachusetts, the college town that he called home for almost all his life. His father, a journalist and author of earnest books and poems on social issues, embraced the inflexible Puritan values of strict Congregationalism. Young Percy, captivated by the unambiguous empiricism of science, rebelled against such blind orthodoxy and ultimately rejected any organized religion, much to the sorrow of his father. Yet he retained the Protestant ethic of honesty, idealism, and hard work throughout his academic and professional life. He applied these traits as a diligent student, both in the public schools of Newton, Massachusetts, and at Harvard University, where he graduated *summa cum laude* in 1904.

Bridgman spent his entire professional career at Harvard, where he rose through the ranks from doctoral student and research fellow, to instructor and junior faculty member, and then on to ever-more-prestigious senior professorial positions, until his final years as Professor Emeritus. Former friends and colleagues remember him as a man of reserved grace and quiet charm. In all of his passionate interests – chess, handball, gardening, photography, mountain climbing, and, most of all, high-pressure research – he found an outlet for his solitary kind of genius.

Bridgman was a loner. He had little patience with committee work, faculty politics, or teaching, and he successfully avoided routine university chores. He was described as a terse, even cryptic lecturer to undergraduates (fig. 11), and he was uncompromising in his demands on students, who rarely lived up to his standards. He was also disappointed in most graduate research students, even though he had the pick of the best young physicists in the country. In the early 1920s John C. Slater, later a leading theoretician of quantum mechanics, and J. Robert Oppenheimer, who was to become the controversial head of the Manhattan Project, were among his trainees. Yet, given a choice, he rebuffed such bright students and worked alone or with his research assistant, Leonard Abbott.

Rain or shine, Bridgman arrived at his physics lab by bicycle. He usually appeared sloppily dressed, often wearing an old rumpled suit and slouched hat that looked as if they'd been slept in. Those who met him in later years, when the Nobel Prize winner had to deflect a steady flow of gawkers and favor-seekers, remember a man fiercely protective of his time and privacy.

When Hatten S. Yoder, Jr., a distinguished high-pressure researcher in his own right, had his first, strange meeting with the legendary figure in the fall of 1946, he found that Bridgman budgeted his time to the minute, with high-pressure experiments dictating when and where he could be seen. As a graduate student in geology at M.I.T., a few miles downstream from Bridgman's Harvard lab, Yoder had called for an appointment to talk about the design of a high-pressure apparatus. Promptly at the appointed hour Bridgman came to his laboratory door, opened it a few inches, and said, "You have seven minutes."

Standing on opposite sides of the partially opened door, Yoder asked the master about pressure vessels, gaskets, and heating elements. Bridgman was cordial in tone and generous with advice, but after seven minutes he abruptly ended the meeting; it was time for another experiment.

Fig. 11 Percy Bridgman lecturing on high-pressure research before the Cornell Section of the American Chemical Society in 1948. (Courtesy of Harvard University, Cruft Photo Lab.)

Yoder was granted several other interviews – on one memorable occasion he spent a full half hour in Bridgman's tiny cubicle office – but they were all no-nonsense meetings. Bridgman would talk only about science with a single-mindedness that many found to be offputting; he had no time to waste on social niceties. Alvin Van Valkenberg, a high-pressure researcher at the National Bureau of Standards, remembered that, outside of a small circle of friends, "everybody was scared of him."

Bridgman was not without social grace when he set his mind to it, and he courted and married Olive Ware in 1912 while he was a junior faculty member. Their personalities evidently complemented each other wonderfully, Percy's self-absorption in the details of laboratory work balanced by Olive's infectious enthusiasm for an active social life. Thanks to Olive, Percy was no stranger to Cambridge's rich intellectual climate of concerts, theater, lectures, and academic dinner parties, though he was known to cancel such engagements on short notice if an experiment demanded his attention.

Two children, born in 1914 and 1915, enlivened the Bridgman household and, by Percy's own admission, at times proved a distraction from laboratory business. Bridgman's sense of duty to the family was strong, but as often as not he interpreted this duty as a mandate for maintaining the highest possible productivity as a researcher. The physicist's frustration at being away from science for any extended period of time tended to argue for a clear division of labor in the Bridgman household.

The Harvard school year, from September through May, was spent primarily in laboratory research, but the family enjoyed summers in the White Mountains in Randolph, New Hampshire, where Bridgman built a vacation home (fig. 12). He devoted much of the summer months to compiling the previous year's research as well as writing a number of books on science and philosophy, but his hours were also occupied with hiking, gardening, and socializing with neighbors in the exclusive vacation community. Yet all of these activities were subordinate to a man who devoted his life to science as single-mindedly as anyone you're ever likely to meet.

* * *

Whatever Percy Bridgman's shortcomings as a conversationalist or socializer, he more than made up for by being one of the greatest experimentalists of his generation. He was a brilliant inventor with excep-

Fig.12 Percy Bridgman, *circa* 1920. (Courtesy of Harvard University, Cruft Photo Lab.)

tional manual dexterity and a thorough practical knowledge of metal-working and machine tools. Much of his most significant work involved the design and construction of advanced pressure seals and gauges. In fact, Bridgman's research career actually began with a lucky discovery in the machine shop.

In 1905, while working on his doctoral thesis – a rather ordinary study on the effects of modest pressure on matter and light – a minor explosion destroyed a critical piece of glassware. Unfortunately, the replacement part would have to be shipped from Europe. Frustrated, Bridgman busied himself modifying an existing high-pressure apparatus he found elsewhere in the lab. In the process he stumbled upon a new design for high-pressure experiments that would become the cornerstone of his career.

Bridgman's extraordinary discovery involved a simple modification in the machinery used to generate high pressure. Every high-pressure experimental system incorporates three principal components: a press, an experimental device, and a sample. The press is simply a frame or support that generates force, usually along a central vertical axis. The press dominates a high-pressure laboratory, often standing taller than a man, and usually has cylindrical steel supports as thick as tree trunks to brace its massive metal end plates. Any of three different mechanisms – screws, weights, or hydraulics – can be employed to provide the compressive force. Presses are usually rated in tons – typically a few hundred tons in Bridgman's day – according to the force that can be exerted on a sample.

Bridgman's antiquated press at Harvard used a giant screw to apply force. It operated something like an antique printing press, a hand-operated wine press, or the vise in your workroom. Although Bridgman had to heft an awkward six-foot-long wrench to turn the central load-applying compressor screw, his apparatus was much less cumbersome than older screw drives. He recalled seeing "enormous capstan-like arrangements that required the force of one or two men to operate" in other high-pressure laboratories.

Heavy metal weights piled on top of a press provided an effective, though even more laborious, alternative means of applying pressure, and some presses were even fitted with giant overhead tanks of mercury liquid to provide continuous variation of the overhead weight. But all things considered, the most convenient presses employed a hydraulic ram to generate large forces along the vertical axis.

The experimental device, which rests between the flat steel plates of the press, is a carefully crafted metal machine that transfers the press's great force to a small sample. The design and construction of a device is complicated by the requirement that it constantly adapts to the size of the sample, which invariably becomes smaller when squeezed. Thus the sample chamber must have at least one moveable wall, with the moving parts sealed against leakage. One of the most common devices in Bridgman's day was a kind of piston-in-cylinder arrangement, in which the sample was placed in a closed cylinder resting on the bottom surface of the press. A snug-fitting metal rod, pressed into the sample chamber from above, applied the pressure. Other devices used a simple pair of circular anvils or a deformable casing of iron or copper to transfer pressure.

Perhaps the most challenging part of squeezing matter is keeping the sample from squirting out; the design of the sample chamber is thus critical to success. Physicists at the beginning of the twentieth century found that many samples were difficult, if not impossible, to confine at high pressure. A central problem was finding a suitable leak-proof seal. Researchers experimented with new gasket materials, coated contact surfaces with sealing compounds made of thick oils or greases, machined metal parts with painstaking care, and plated parts to achieve an even better fit. Even so, the samples – whether gases, liquids, or solids – often squirted out the sample chambers. As a result, experimentalists at the turn of the century rarely achieved sustained pressures of more than 2000 or 3000 atmospheres before their sample chambers failed.

However, by some combination of luck, intuition, and keen observation, Bridgman stumbled upon both a wonderful new experimental design (fig. 13) and the perfect material with which to execute it. He used a very soft solid as the pressure seal and kept this soft solid at a pressure always a bit higher than that of the sample. In this way, the pressurized sample could not possibly leak past the higher pressure of the seal.

To accomplish this feat, Bridgman substituted an easy-to-assemble two-piece piston for the traditional single metal piece. He inserted a short metal piston segment that looked something like an inverted mushroom with a flat disk-shaped head and short cylindrical stem into

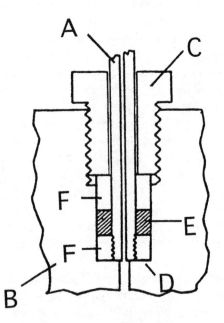

Fig. 13 Bridgman's unsupported-area packing represented a new kind of high-pressure connection between a pipe and pressure vessel. The area of the gasket seal is less than the area of the sample, so the pressure on the seal is always greater, preventing sample leaks. This figure, from Bridgman's *The Physics of High Pressure*, provided an enlarged view of the connection. (From the Author.)

the cylinder, flat end first. Around the mushroom's stem Bridgman placed a lifesaver-like ring of Sioux Indian pipestone, a soft, carvable rock. He then placed a second piston around the first. The second piston was a carefully machined cylinder with a central hole the same diameter as the mushroom stem; its shaft was slightly longer, to accommodate the stem with room to spare. The beauty of Bridgman's device lies in its exploitation of pressure as a force acting on an area. Regardless of the force applied to the piston, the ring-shaped pipestone gasket has less area and therefore is subjected to more pressure than the circular sample of identical diameter.[4]

Bridgman later recalled the crucial discovery: "In the interval of waiting for the replacement [part for my thesis work] I tried to make other use of my apparatus for generating pressure. While designing a closure for a pressure vessel, so that it could be rapidly assembled or taken apart, I saw that the design hit upon did more than originally intended; the vessel automatically became tighter when pressure was increased, so that there was no reason why it should ever leak."[5]

This discovery of an impervious, self-tightening seal transformed high-pressure research. "The whole high-pressure field opened up almost at once before me, like a vision of a promised land, with the discovery of the unsupported area principle of packing, by which the only limit to the pressures attainable was the strength of the metal parts of the apparatus."[6] The original thesis project was abandoned, and Percy Bridgman took off on a lifetime of exploration in high pressure.

Within a couple of years he could routinely attain pressures of 7000 atmospheres, or about 50 tons per square inch – considerably higher than that achieved in any other laboratory, even though Bridgman was hampered by his antiquated screw compressor. Rather than immediately strive for even higher pressures with a modified press, however, Bridgman focused on a less exotic problem. He realized that attaining high pressures meant little without the means to measure those pressures. He therefore spent much of his graduate student years devising new ways to measure the unprecedented pressures he was now able to achieve.

Scientists have created a bewildering variety of units to quantify the phenomenon of pressure. Pressure is defined as a force applied to an area, such as pounds per square inch. One atmosphere (the unit I'll use in most of this book) is produced by a 14.7–pound weight on an area of one square inch; one atmosphere is also, by pure chance, almost exactly the metric pressure generated when one kilogram pushes down on an area of a square centimeter. The bar, another common pressure unit, is defined as one million dynes (a unit of force) per square centimeter. Fortunately, that pressure is only slightly less than one atmosphere (1 bar = 0.987 atmospheres), so bars and atmospheres are more or less interchangeable.

Meteorologists, who often measure atmospheric pressure with a mercury-filled barometer, talk about inches of mercury; the weight of all the air above you is more or less equal to the weight of a liquid mercury column thirty inches high. This barometric pressure changes slightly depending on whether your region is experiencing a high- or a low-pressure weather system. To make matters even more complicated, physicists decided a few years ago that the official unit of pressure should be pascals, defined as a newton (another unit of force) per square meter – about equal to the force of a feather resting on your finger. It takes exactly 10 000 pascals, or 10 kilopascals, to make a bar, which, as you remember, is close to an atmosphere. I could also tell you about torrs and slugs, but you don't really want to know.[7]

Bridgman's first step was to establish a primary pressure scale, based only on a direct measurement of force per area. In the nineteenth century, for pressures up to a few hundred atmospheres, scientists used tall open columns of mercury to do the trick. In France, for example, researchers used the Eiffel Tower to support a mercury column several hundred feet tall (one atmosphere is equal to about thirty inches of mercury), while even higher columns were constructed in a deep French coal mine.[8] But at the pressures of Bridgman's new device, the open mercury column would have to be several miles tall. Obviously, another approach was required. Bridgman based his primary pressure measurement on an accurate set of metal weights (the force) pushing down on a precisely machined piston of known cross-sectional area. In such an assembly, the pressure – the force acting on an area – could be measured directly from the weight pushing on the piston.

In principle, all of his experiments could have employed such a pressure measuring system, but handling heavy weights and insuring precise piston areas can be a tricky business. It proved much easier to use the primary weight scale in just one crucial set of measurements in order to calibrate a more convenient secondary scale – a scale based on the way an easily measured material property varies with pressure, the same way we commonly use the volume of mercury as a measure of temperature in a thermometer. Many researchers had relied on a fluid's volume change – its compressibility – to measure pressure, while others calibrated pressure by observing the elastic deformation of a metal bar or spring. Bridgman settled on a different approach, based on the significant decrease of mercury's electrical resistance with pressure. His mercury resistance gauge would provide a fast and reliable method to measure pressure for decades of experiments to come.

The results of Bridgman's thesis, published as two papers in the 1909 Proceedings of the American Academy of Arts and Sciences under the collective title, "The Measurement of High Hydrostatic Pressures," presented a comprehensive protocol for attaining and calibrating unprecedented pressures in the laboratory.9 Yet, in spite of fifty pages of experimental details, Bridgman did not divulge the geometry of his double-piston apparatus or the identity of his new pipestone packing material. He knew he had the edge on the competition and, for a time at least, he alone would profit from that discovery.

* * *

Armed with new techniques to confine and calibrate pressure, Bridgman was ready to set a few world pressure records. As a research fellow in 1910, he replaced the antiquated screw compressor with a hydraulic ram press and immediately achieved more than 20 000 atmospheres, a pressure so high that he was almost apologetic about claiming it. "The magnitude of fluid pressure mentioned here requires brief comment, because without a word of explanation it may seem so large as to cast discredit on the accuracy of all the data."10

Within a few short years Percy Bridgman had created a laboratory like none other in the world. He had at his disposal the apparatus to squeeze samples as they had never been squeezed before, and the logical research strategy was obvious to him. He decided "to use this new technique to the limit, attacking with it any problems in which the information to be expected from the behavior under high pressures seemed likely to be of significance."11

He then engaged in an unabashed program of experimental prospecting, pressurizing almost anything he could lay his hands on, confident that a world of fascinating phenomena lay in wait. He acknowledged that "this is not the usual procedure in scientific work, in which the problem usually presents itself, and the suitable technique discovered."[11] But he proceeded without apology to make wonderful discoveries, and in the process also served as a model for countless thousands of researchers to follow – scientists who, in an age of exorbitantly expensive research apparatus, become tied for life to a piece of fancy hardware. When scientists spend a million dollars for an electron microscope, or a group invests several billion dollars for a particle accelerator, it becomes imperative for them to select experiments that require that piece of equipment.

That's not necessarily a bad situation, especially if you can make discoveries like Percy Bridgman's. After some fairly routine studies of metal compressibilities, he made his first high-pressure headlines by squeezing water to more than 20 000 atmospheres – five times more than any previous investigation. At such pressures, Bridgman declared, "results for water are much more varied and richer . . . appearing in no less than five [solid] forms."[12] In 120 pages of details, he described the ranges of temperatures and pressures at which the extraordinary and unexpected forms of high-pressure ice occur – each variety heralded by a sudden change in sample volume.

One of these varieties, Ice VI, stable to at least 95°C, or 200°F, was dubbed "hot ice" in the popular media, and evoked a flurry of ill-founded interest in commercial applications. E. F. McPike, manager of an Illinois Central Railroad fruit-shipping concern wrote to Bridgman, "I have read a brief note in the newspaper regarding your production of a solidified form of hot water. Might we inquire if in your opinion this process has commercial possibilities for use in transit in a manner similar to ice to protect fruits and vegetables . . . against cold instead of heat? Portable heaters have been used to some extent by carriers and if there is any prospect of a better process we would be glad to know about it."[13] Bridgman had to inform Mr. McPike that Ice VI, like all the other high-pressure forms, can only exist under pressure and melts immediately upon its release.

Following the fascinating water work of 1911, Bridgman tackled other liquids. In 1912 he studied a dozen common organic fluids, including alcohols, ether, and acetone, and in the next year expanded his domain to include more and more materials – solids, liquids, and gases. In 1914 he squeezed a dozen common salts and several pure elements including sodium, potassium, phosphorus, and mercury.

For the next twenty years he studied the compressibility and electrical properties of hundreds of compounds to 20 000 atmospheres. Every experiment was virgin territory, and time after time he uncovered new crystalline structures that can only form at high pressure – just like diamonds. But apart from a few high-pressure electrical resistance measurements on graphite in the early 1920s, however, Bridgman did not publish any data on the forms of carbon, nor did he record any concerted effort to make diamonds in those early days. But it would have been strange indeed if he had not, on occasion, placed a small disk of graphite in his mighty press, just to see what happened.

* * *

For two decades Bridgman was content to study matter at pressures to 20 000 atmospheres, but higher pressures beckoned. Fascinating physics of matter waited to be explored at higher pressures, and throughout the later stages of his career, from the mid-1930s on, the design and improvement of high-pressure gear became a recurrent theme.

Bridgman suspected that there were probably limits to the pressures humans can achieve in the laboratory, for even the best built experimental apparatus has a limit to the amount of stress it can endure before breaking. If you want more pressure, you had better be prepared to break a lot of equipment. High-pressure researchers often refer to their sealed cylinders, especially those pressurized with gas, as "bombs" – a term that is quite realistic in its bravado.

Every high-pressure experiment carries with it the danger of catastrophic failure, and rare is the worker in the field who does not duck flying metal at some point in his or her career. The most serious high-pressure accident at Harvard occurred on May 19, 1922, when engineering research fellow Atherton K. Dunbar and an assistant William Connell were killed while pressurizing a tank with oxygen. A terrible explosion ensued, as described in the following account: "The middle section of the basement was completely wrecked. Dunbar was blown to pieces, and the carpenter Mr. Connell was instantly killed. Eight students in the room above were injured, the floor being lifted bodily, and a heavy dynamo turned over onto some of them."[14] Shortly thereafter, Harvard's high-pressure research was moved to a sturdy concrete and wood garage that had previously served the campus ROTC unit. The new research facility, a bit crude but eminently functional, was chris-

tened the Dunbar Laboratory. For almost forty years it was the site of
Percy Bridgman's experimental triumphs.

Bridgman was not injured, nor were any of his presses involved in
the Dunbar tragedy, but he braved more than his fair share of breakage,
and devoted an entire chapter of his 1931 book *The Physics of High
Pressure* to the "variety of interesting ways" that a pressure vessel might
rupture.[15] Pressure systems, like chains, fail at their weakest point.
Bridgman found that pistons made of tough file steels or ball-bearing
steels (fig. 14) were sufficiently strong as they compressed a sample, but
the steel cylinders, which bulged out under tension, frequently cracked
and exploded. Bridgman realized that this problem was not unique to
pressure research. The same kind of failure concerned the military,
whose cylinders discharged bullets and explosive shells. To compensate
for such potential breakage, he adapted a clever method for shrink-
fitting a thick steel girdle around the central cylinder to compress the
tube and reduce the chance of high-pressure failure.

Nevertheless, laboratory explosions were inevitable, and they could
lead to curious superstitions. In experiments with sealed tubes, failure
often occurred when one end of the pressure vessel sheared off. The
resulting explosion could launch the contents, including the sample
and metal rods used to pack the sample chamber, at the speed of a rifle
bullet. Once one of Bridgman's Dunbar Laboratory bombs blew out and
embedded a steel filler rod deep in the adjacent wall a few feet away.
Bridgman felt it was a lucky miss, for he often walked by that very spot
on the way to his cubby hole office (fig 15). After the incident, it is said, he
would carefully step across the spot. By the 1950s, after decades of work,
Bridgman's lab. walls had quite a few impressive holes, but tradition
has it that he never had them repaired.

Every high-pressure experiment is limited by the strength of the
materials involved; every material will break when stressed beyond its
limits. Even the strongest steels – even diamonds – will break if pushed
too far. Designers of high-pressure apparatus thus focused on two
major concerns: selecting the strongest available materials, and finding
the most dependable way to build a device out of them. These problems
were never far from Bridgman's mind, and he jumped at the chance to
build better equipment as stronger, more advanced steels became available.

His simplest device, which is still used today in many laboratories,
was nothing more than a vise with opposed tapered anvils. The first

Fig. 14 Bridgman's high-pressure "bomb," *circa* 1950, consisted of a vertical hydraulic piston that was pressed into a cylinder. (Courtesy of H. S. Yoder, Jr.)

Fig. 15 Harvard University's Dunbar Laboratory, *circa* 1950, home of Percy Bridgman's high-pressure lab, was converted from an ROTC garage. (Courtesy of H. S. Yoder, Jr.)

model, introduced in 1935, employed anvils of hardened steel with round, flat faces a quarter-inch across (fig. 16).[16] The opposed anvils could be rammed together to generate 50 000 atmospheres of pressure on the sample, which was surrounded and confined by a ring of pipe-stone. The device was relatively easy to use and was soon dubbed the "simple squeezer". Within a few years he had compressed more than 100 elements and compounds to this extreme pressure in epic scientific papers with titles like "Polymorphic transitions of 35 substances to 50 000 kg/cm²" (1937)[17] and "The compression of 46 substances to 50 000 kg/cm²" (1940)[18] – titles that speak of an exhaustive research effort.

In the late 1930s Bridgman learned of a new material, a variety of tungsten carbide sold by General Electric's Carboloy Division – a

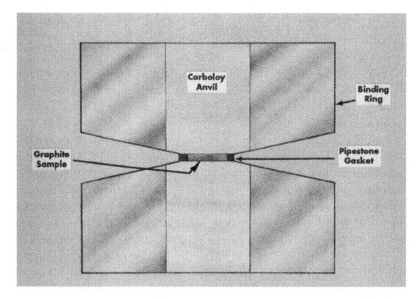

Fig. 16 Bridgman's simple opposed-anvil device featured two tapered anvils that clamped together like a vise. A ring-shaped gasket confined a sample, which could be squeezed to more than 100 000 atmospheres. (Courtesy of F. R. Boyd.)

material with hardness and stiffness much greater than ordinary steels. Bridgman lost no time in the construction of new carbide anvils. He supported the anvils with shrink-fit girdles of steel and placed them in his press. Almost at once he achieved a useful pressure he believed to be in excess of 100 000 atmospheres (we now know it was closer to 70 000 atmospheres – still a record for his time). Armed with such extraordinary squeezing power, Bridgman commenced a period of incredible productivity. "Modest" pressures below 50 000 atmospheres were a cinch, and led to mammoth surveys like "Rough compressions of 177 substances to 40 000 kg/cm^2" completed in 1948.[19] During the 1947–48 academic year Percy Bridgman averaged almost one new sample every day, six days a week.

Part of Bridgman's astonishing productivity was a result of his machining skill, which allowed him to build and repair high-pressure devices himself. He often made his own steel pressure vessels even though he had a full time machinist, Charles Chase, working with him. His prowess and patience with a lathe are legendary. In one publication he described the extraordinary efforts he made to drill his own high-pressure tubing, taking extreme care to start with high-grade metal, centering the drill, and rotating the tubing at high speed while keeping the growing hole clean. "It is easy, if all precautions are observed, to drill a hole 1/16 of an inch in diameter 17 inches long in from seven to eight hours."[20]

* * *

As the world's foremost authority on high pressure, and master of the world's foremost array of high-pressure equipment, Percy Bridgman was no stranger to the idea of making diamonds. In the November 1955 issue of *Scientific American* he reminisced: "I suppose that over the last 25 years an average of two or three people a year have come into my office, offering to share the secret and the profit of making diamonds in return for my constructing the apparatus and reducing the idea to practice. The problem has got into the thriller literature, and I have often encountered the belief that the successful solver of this problem would be in danger of his life from the Diamond Syndicate."[21]

The professor was hardly immune from the lure of transmuting graphite into diamond. David T. Griggs, among the only Ph.D. candidates to work directly under Bridgman, wrote "It was my privilege to work in Bridgman's laboratory during the period when working pressures were increased from 20 000 to 100 000 [atmospheres]. As each new apparatus was readied for trial, I noticed that Bridgman would become secretive and brusque. During the first run, visitors were not welcome. I subsequently learned that in each case *graphite* was the first substance tried."[22]

Bridgman's strategy for making diamonds was simple. Squeeze a sample of graphite as hard as possible between two strong surfaces. The hardest steels crafted into his simple squeezer splintered at about 60 000 atmospheres without any change in the graphite, so Bridgman turned to his new superhard Carboloy anvils. By replacing steel components with Carboloy, he could subject graphite to much higher pressures. The opposed carbide anvils provided an estimated 150 000 atmospheres before they broke – still not enough to transform diamond. But Bridgman had not run out of ideas. Next he tried a single carbide anvil, less than a tenth of an inch across, crushed into a flat carbide surface. Bridgman reported pressures estimated at an incredible 425 000 atmospheres – about 3000 tons per square inch – yet graphite persisted, bouncing right back to its original form when the pressure was released. Bridgman abandoned that effort to make diamond with the wry observation that in graphite he had discovered "nature's best spring."

Bridgman was not through with the diamond-making game, however. General Electric and its Carboloy Division, the principal suppliers of the carbide components that could only be shaped with diamond-impregnated tools, and the Norton and Carborundum Companies, the largest consumers of industrial diamond, joined forces in January of 1941 and persuaded Percy Bridgman to spearhead another diamond synthesis effort. They installed a brand new thousand-ton hydraulic press at the Harvard Dunbar Lab, provided a staff of assistants to help build and operate experimental devices of Bridgman's design, and committed resources for a five-year research and development program.

It was well known that at room pressure and very high temperatures (above 1500°C in the absence of oxygen) diamond reverts to graphite. Perhaps, Bridgman and others thought, the reverse would happen at high pressure and similar temperatures. The principal roadblock was that steel or carbide anvils soften at high temperature, thus greatly reducing the attainable pressure. In one set of experiments Bridgman subjected graphite mixed with tiny diamond seed crystals to red heat, perhaps 700°C, at 75 000 atmospheres. Nothing happened.

In a second series of experiments Bridgman heated wafers of graphite outside a press to 2800°C and quickly transferred them while incandescent to the press. The seven-second operation proved too slow, the graphite cooled too quickly, and again no diamonds formed, though they did manage to destroy all traces of the original diamond seeds, which had converted to graphite. Unbeknownst to Bridgman, P. L. Gunther and his German colleagues had tried the exact same approach in the early 1940s, and they too were unsuccessful.

After those failures, Bridgman and coworkers developed yet another procedure by surrounding a disk-shaped sample of graphite and diamond seed crystals with thermite, a chemical that burns at very high temperature. They pressurized the sample assembly between carbide anvils and then triggered a fire, thereby creating sustained conditions of 30 000 atmospheres and 3000°C for a few moments. Once again, no diamonds formed, but for the first time in an experiment at such high temperatures the diamond seed crystals did not revert to graphite, suggesting that the investigators were on the right track.

The principal conclusion of the unsuccessful research was that sustained temperatures greater than 1000°C at pressures above 50 000
atmospheres – conditions beyond the Harvard lab's capabilities –
would be required to make diamonds. Bridgman had invested only two
years in diamond making when he was diverted by the demands of
other World War II research. In one project he was called upon to
measure the compressibilities of uranium and plutonium – data critical
to the Manhattan project. The diamond-making effort had to be abandoned.

In 1946 Percy Bridgman was honored with the Nobel Prize in
physics "for the invention of an apparatus to produce extremely high
pressures, and for the discoveries he made therewith in the field of
high-pressure physics." Four decades of achievements had placed him
in the foremost ranks of scientists. He had overcome obstacle after
obstacle, broken record after record, and single-handedly transformed
the science of pressure. Even so, the mystery of diamond making had
eluded him.

* * *

In the spring of 1959 scientists at the National Bureau of Standards
approached Harvard University Press and proposed that the experimental papers of Percy Bridgman be collected in one set of volumes.
Many of his more than 200 papers, including classic descriptions of
high-pressure apparatus and studies of hundreds of materials, were
scattered in dozens of different periodicals, many of which were
unavailable in smaller libraries. Bridgman agreed to collate and edit the
volumes himself, while the National Science Foundation supported the
project with a grant in July, 1961.

Bridgman selected 198 papers for inclusion in the seven-volume set,
and prepared a preface, annotations, and indices for the work. It was to
be his last scientific effort. In the spring of 1961 he was increasingly
troubled by what was at first thought to be muscular rheumatism. The
correct diagnosis, an inoperable bone cancer known as Paget's disease,
was made in midsummer. Suffering from ever increasing pain and loss
of muscle control, Bridgman worked on the collected papers project to
the very last day, when the indices to his life's work were completed and
sent to the publisher. Then, determined to control his own destiny,
Percy Bridgman went to the pumphouse behind his summer home,

sawed off the barrel of a shotgun, placed it in his mouth, and pulled the trigger. Like the careful experimenter that he was, he left nothing to chance.

Though he gave no formal classes and taught few students, Bridgman left an incalculable legacy to solid-state physicists and high-pressure researchers. His colleagues mourned the loss not only of a friend and associate, but also of his tremendous knowledge, much of it gained through years in the laboratory and never written down, knowledge that died with him.

Some saw in Percy Bridgman's last violent act a final affirmation of his intellectual honesty and integrity – an effort to spare friends and family the pain and expense of a protracted decline. Others condemned the suicide as a thoughtless cruelty to his surviving wife and children. Who can imagine what passed through his mind, or what his pain must have been as he wrote his last words: "It isn't decent for Society to make a man do this thing himself. Probably this is the last day I will be able to do it myself."

Notes

1. As quoted in: Maila L. Walter, *Science and Cultural Crisis: An Intellectual Biography of Percy Williams Bridgman (1882–1962)* (Stanford, CA: Stanford University Press, 1990), p.25.
2. Most of Bridgman's published experimental papers were collected and republished in a seven-volume set shortly after his death: Percy W. Bridgman, *Collected Experimental Papers* (Cambridge: Harvard University Press, 1964). Additional non-technical writings appear in: Percy W. Bridgman, *Recollections of a Physicist* (New York: Philosophical Library, 1950).
3. Maila Walters's biography, *Science and Cultural Crisis*, op. cit., is the principal source of background information for this chapter. Additional personal details are found in: James B. Conant and others, *Percy Williams Bridgman, 1882–1961* (Cambridge: Harvard University, Department of Physics, 1961). See also the biographical essay by Edwin C. Kemble, Francis Birch, and Gerald Holton in: *Dictionary of Scientific Biography*, vol. 2 (New York: Scribners's Sons, 1970), pp. 457–461.
4. Bridgman's own illustrated description of high-pressure packings appears in: P. W. Bridgman, *The Physics of High Pressure* (New York: MacMillan, 1931), pp. 30–40.
5. P. W. Bridgman, "Recent work in the field of high pressures." *American Scientist* **31**, 10 (1943).
6. Bridgman, *The Physics of High Pressure*, p. 78 (see note 4).
7. The torr, named for the Florentine physicist Evangelista Torricelli (1608–1647), is defined as the pressure generated by a column of mercury 1 mm tall at 0°C – a low pressure, somewhat less than a thousandth of an atmosphere. For historical reasons

that are unclear to my colleagues and me, the torr is still almost universally used in vacuum gauges on all sorts of experimental apparatus. The best laboratory vacuum pumps produce pressures as low as 10^{-7} torr.

The slug, an archaic mass unit from the British system of weights and measures, is defined as the mass that, when acted upon by a force of one pound, will experience an acceleration of one foot per second² (approximately 14.6 kilograms). Thus, a British unit for pressure was one slug per foot², or about 0.015 atmosphere.

8. A history of pressure calibration through Bridgman's developments appears in: Bridgman, *The Physics of High Pressure*, pp. 60–77 (see note 4).

9. P. W. Bridgman, "The measurement of high hydrostatic pressure, I. A simple primary gauge." *Proceedings of the American Academy of Sciences* **44**, 201–217 (1909); "The measurement of high hydrostatic pressure, II. A secondary mercury resistance gauge." *Proceedings of the American Academy of Sciences* **44**, 221–251 (1909).

10. P. W. Bridgman, "The measurement of hydrostatic pressures up to 20 000 kilograms per square centimeter." *Proceedings of the American Academy of Sciences* **47**, 321–343 (1911).

11. P. W. Bridgman to P. J. Risdon, Dec. 10, 1922, as quoted in Walter's, *Science and Cultural Crises*, p.33.

12. P. W. Bridgman, "Water, in the liquid and five solid forms, under pressure." *Proceedings of the American Academy of Sciences* **47**, 441–558 (1911).

13. McPike to Bridgman, Aug. 24, 1912, as quoted in Walters, *Science and Cultural Crisis*, p. 24 (See note 1).

14. As quoted in: Ibid., pp. 37–38.

15. Bridgman, *The Physics of High Pressure*, p. 78.

16. P. W. Bridgman, "Polymorphism, principally of the elements, up to 50,000 kg/cm²." *Physics Review* **48**, 893–906 (1935).

17. P. W. Bridgman, "Polymorphic transitions of 35 substances to 50,000 kg/cm²." *Proceedings of the American Academy of Sciences* **72**, 45–136 (1937).

18. P. W. Bridgman, "The compression of 46 substances to 50,000 kg/cm²." *Proceedings of the American Academy of Sciences* **74**, 21–51 (1940).

19. P. W. Bridgman, "Rough compressions of 177 substances to 40,000 kg/cm²." *Proceedings of the American Academy of Sciences* **76**, 71–87 (1948).

20. P. W. Bridgman, "The technique of high pressure experimenting." *Proceedings of the American Academy of Sciences* **49** (1914), p. 638. For additional details see: Bridgman, *Physics of High Pressure*, p. 41.

21. P. W. Bridgman, "Synthetic diamonds." *Scientific American* **193**, 42–46 (November 1955). Bridgman's experiments are detailed in: P. W. Bridgman, "An experimental contribution to the problem of diamond synthesis." *J. Chem. Phys.* **15**, 92–98 (1947).

22. As quoted in: H. Tracy Hall, "The transformation of graphite into diamond." *American Association for Crystal Growth Newsletter* **16**, 2–4 (1986), p.2.

CHAPTER 4

BALTZAR VON PLATEN
AND THE INCREDIBLE DIAMOND
MACHINE

As I passed the School of Botany, I saw that one wall was covered with Virginia Creeper. Its leaves were a beautiful red. Every autumn brought the change from green to red, and passers-by would pause at the display of colour in sudden admiration. I was one of these passers-by, and had no idea that my pleasure in this seasonal beauty would later play an important role in my life, . . . I had no suspicion that later this sight would show me the way to construct a machine which makes diamonds.[1]

BALTZAR VON PLATEN, *Modern Very High Pressure Techniques*, 1962

O F ALL THE MANY ATTEMPTS TO create diamonds in a laboratory, none was more outrageously impractical and shamelessly extravagant than the effort of ASEA, Sweden's major electrical company. But, then again, the Swedish company's diamond project was conceived and constructed by Baltzar von Platen, a scientist who was, by some accounts, quite mad.

In the early 1940s two rival companies had taken up the diamond-making challenge. It was more than coincidence that both groups were principally concerned with electrical power, appliances, and light. Both companies relied on diamond tools to machine great dynamos and other electrical devices. Both had considerable expertise in using electricity to generate the high temperatures critical to success in the venture. And both expected to win the race. The two companies were Allmanna Svenska Elektriska Aktiebolaget, or ASEA (roughly translated, Swedish General Electric Company), and the General Electric Company in the United States. Of the two, ASEA always seemed to be one step ahead.[2]

The ASEA effort started with the dreams of one man. Baltzar von Platen (fig. 17)–called "a genius maniac" by some – was the kind of half-

Fig. 17 Baltzar von Platen (left) and Erik Lundblad, *circa* 1955. (Courtesy of Erik Lundblad.)

crazy inventor stereotyped in popular movies and dime-store novels. Tracy Hall, a key player in the General Electric diamond-making effort, first met von Platen in 1957 during a visit to Stockholm. "He drove an ancient car," Hall recalls. "When he stopped at a red light he'd turn the engine off. Then he'd restart it when the light turned green." Longtime associate Erik Lundblad was well acquainted with von Platen's quirks. "He carried a name well known in Sweden since the eighteenth century, when his namesake built a great canal from the Baltic Sea to the North Sea and was raised by the King to the nobility. Everyone believed that our Baltzar was [a descendent] of the old man and he himself spared no pain in letting people believe they were right – which they were not."

Some colleagues speculated that his idiosyncracies may have arisen, in part, from a facial deformity – a large disfiguring birthmark on his left chin that embarrassed him greatly. But whatever the cause, he marched to a very different drummer. Baltzar von Platen's most famous and curious invention, a portable thermal refrigerator that produced ice with a gas flame, epitomized his quirky brilliance. He used an ammonia refrigerant that was vaporized by a blue-hot gas flame at one end of his contraption; the ammonia gas flowed through the system to condense in refrigeration coils at the other end. Before the universal availability of electricity, von Platen's thermal refrigerators, manufactured and marketed by Electrolux, were a godsend that transformed rural life in many parts of Europe and America.

This paradoxical appliance was the height of sanity compared to von Platen's other efforts. With his reputation as an inventor assured, he went on to play the role with gusto. "His imagination was unlimited," Lunblad recalls. "His destiny was to solve the world energy problem by designing a perpetual motion system. To make the idea more reliable he declared the second law of thermodynamics . . . invalid." Evidently, a lot of people believed him. Major corporations, including Volvo, sponsored his research at considerable expenditure. "Once they realized they had been swindled they often felt so embarrassed they never opened any law suits."

In this freewheeling spirit von Platen designed his incredible high-pressure, diamond-making system. He began thinking about synthetic diamonds as early as 1930, when he read an article describing the extreme conditions required to convert graphite to diamonds. He knew that the tremendous temperatures and pressures necessary to make diamonds were too much for any device made of steel. He concluded that to make diamonds you had to accept as inevitable the destruction of your machine in the process. But, he emphasized, if you accept the fact that your machine's destruction is inevitable, that knowledge allows you to approach the design in a radical new way: build the device as if it were already broken!

Many years after the ASEA effort, Baltzar von Platen contributed a strange, rambling chapter on his diamond-making machine to the 1962 monograph *Modern Very High Pressure Techniques*, which is for the most part a rather dry, technical book. However, in prose atypical of scientific exposition, von Platen revealed the mystical source of his inspiration, which was found in a bit of ancient mythology regarding the origin of brilliantly colored autumnal foliage. According to the myth, the spirit-philosopher Demiurge convinced leaves to begin the inevitable process of death and disintegration while still attached to their tree, rather than succumb to death by falling to the ground in a fresh and green state. By beginning the irreversible decay process before falling from the branches, leaves provide some beauty and benefit for others and, in the process, cheat death out of two or three weeks.

This quaint story evidently came forcefully to mind as Baltzar von Platen contemplated the diamond problem. "All at once a thought struck me and an association sprang to life. It was utterly unexpected, for up to then I had done no work at all on the diamond problem, though I had often been tempted to get to grips with it. Suddenly I saw how Demiurge's principle for prolonging the life of the leaves could be applied to a machine for making diamonds. One had merely to convert botanical facts into mechanical ones, and the parallel between the corresponding details seemed to me quite complete."[3]

In von Platen's extraordinary mind, the myth contained the philosophy around which to build an experiment: If destruction is inevitable, then turn that destruction to your advantage. "You know your machine must fall to pieces when it is destroyed – when it is killed – by the enormous pressure. And the form of these pieces will be wrong, since the dead steel has not been invested with . . . wit and knowledge. . . . You must let your machine go to meet Death by dividing it up into pieces, but you must give these pieces the shape that the machine itself would have wished, had it been a living organism like the leaves. Then it will stand much higher pressures, and live longer. And then you will be able to make diamonds."[4]

This *is not* typical experimental protocol, and von Platen's metaphors do not speak persuasively to most working scientists. Nevertheless, although it is difficult to see how a high-pressure invention could follow from Demiurge's principle, von Platen's apparatus successfully generated extraordinary temperatures and pressures, and his idea continues to serve as the basis for many of today's most successful high-pressure machines.

* * *

Von Platen, who enjoyed a steady royalty income from Electrolux for the thermal refrigerator, established his research laboratory in downtown Stockholm in a magnificent early seventeenth-century hunting palace built by King Gustav Adolphus II for his mistress Ebba Brahe. The structure had largely fallen into disrepair, but it provided ample room for research and living quarters. He called this grand, decaying palace his home when he began to construct his diamond-making machine in the late 1930s. The task soon proved too big – physically and financially – for his personal resources, however, so he approached ASEA about his diamond synthesis scheme in 1941. His seductive proposal called for nothing less than the synthesis of gem-quality diamonds several centimeters across and dozens of carats in size. According to Erik Lundblad, von Platen planned to make "home made 'Koh-i-noors,' nothing less would do."

Von Platen's diamond-making strategy was more than a little odd, but given his previous spectacular success, people were reluctant to dismiss these new ideas. ASEA believed the considerable risk of failure was more than offset by the potential bonanza if diamond growing proved successful. By mid-1942, a contract for joint research and development was signed, and designs for new equipment were begun almost at once.[5] Von Platen was to provide whatever high-pressure apparatus he had already built and prepare schematic drawings for additional components of his elaborate experimental device – a pressure vessel built as if it were already broken into pieces. Once these blueprints were finished, the inventor advised on the construction and use of his machine, but his primary attention shifted to other projects.

The hard work of assembling and running the equipment was left in the able hands of ASEA engineers, led by Ragner Liljeblad, head of the company's research and development effort. In 1945, just after World War II, ASEA moved von Platen's pressure equipment to the main corporate laboratory in Västerås, about 100 kilometers west of Stockholm, where engineers had facilities for the necessary heavy-duty machining.

The initial stages of the project required ASEA to build a massive press, or "yoke," capable of sustaining 12 000 tons of force. Into the jaws of the powerful ASEA yoke the team of engineers and technicians inserted the most costly and cumbersome device ever conceived and built for high-pressure research. Von Platen had designed a massive and complex contraption to compress a spherical carbon-rich sample no larger than a marble. The ASEA workers surrounded this sample with a sphere of thermite, an unstable, sometimes explosive chemical mixture of barium peroxide and magnesium metal; once ignited by an electric current, thermite burns at more than 4000°F. The tennis-ball-sized thermite mass was cautiously packed in a shell of soapstone insulation, which in turn was encased by a cube-shaped cover of soft metal, usually copper or iron. The final sample-holding cube (fig. 18) measured about three inches across and weighed a few pounds.

Von Platen's experimental device was designed to apply the same kind of pressure that formed diamond in nature. Deep within the earth,

Fig. 18 A copper-jacketed cube-shaped sample assembly, approximately three inches along each edge, formed the center of the ASEA device. Inside the deformable metal cube was a carbon-rich sample, surrounded by a tennis-ball-sized sphere of thermite. (Courtesy of Erik Lundblad.)

pressure is generated by gravity, which causes rock to squeeze together from all sides at once, much the way a diver is squeezed from all sides by water. This kind of pressure – hydrostatic pressure – is distinct from the "uniaxial" action of a vise, which only compresses along one axis. Von Platen achieved hydrostatic pressure by surrounding his sample with six pyramid-shaped anvils (fig. 19) that came together to form a planet-earth-like solid sphere (fig. 20). By pushing simultaneously on all six anvils, each of which pressed on a different cube face of the sample, tremendous uniform pressure could be achieved at the sample core.

When properly positioned, the six anvils and sample assembly formed an iron ball almost two feet across. The Swedish workers enclosed this entire sphere in a strong copper jacket, creating a high-pressure chamber that weighed in at almost half a ton. Yet this extraordinarily complex arrangement, constructed from more than two dozen

Fig. 19 Six steel and carbide anvils, each a segment of a sphere, compressed the ASEA cube sample. (Courtesy of Erik Lundblad.)

separate pieces, did absolutely nothing by itself. Somehow, the entire device had to be pressurized.

To make diamonds, ASEA technicians sealed the spherical high-pressure chamber into a water-filled cylindrical tank (fig. 21) that would be pressurized to about 6000 atmospheres – a pressure several times greater than that experienced at the ocean's deepest trench. The principal technical challenge was to confine a large volume of water at such

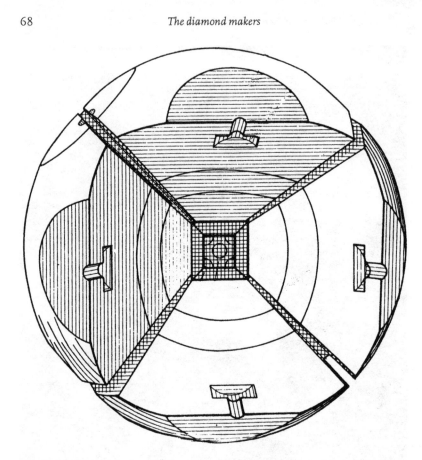

Fig. 20 Baltzar von Platen's split-sphere apparatus incorporated six wedge-shaped anvils, each directed at a face of a cube-shaped sample. This figure illustrates the assembly, with one of the anvils removed. The entire split-sphere assembly was jacketed in copper and immersed in a high-pressure water tank. (Courtesy of Erik Lundblad.)

high pressures; the tiniest leak could create a high-velocity water jet capable of drilling right through a hand.

The researchers wrapped the body of the cylindrical tank again and again with hundreds of loops of taut piano wire to prevent the sides of the pressurized tank from bursting. Then they reinforced the tank's top and bottom by placing it inside the steel yoke, a structure that was more than nine feet tall. The yoke itself was constructed from thick steel plates, welded together and wrapped with even more piano wire. The tank and yoke were hulking, massive chunks of metal, difficult and expensive to assemble. Even so, each experiment placed such tremen-

Fig. 21 A watertight copper jacket surrounded the ASEA split-sphere device, which was almost two feet in diameter. (Courtesy of Erik Lundblad.)

dous stresses on these components that metal fatigue often ruined new tank and yoke assemblies – in some cases after only two or three runs.

Von Platen's experimental hardware was extraordinarily difficult to build, but its operational strategy was simple. Since pressure equals force divided by area, the 6000 atmospheres of water pressure applied to the split sphere's very large outside area would become focused and concentrated to tremendous pressures – greater than 50 000 atmospheres, it was hoped – on the relatively small area of the central cube-shaped sample assembly. As water pressure crushed the sphere and forced the six anvils against the sample assembly, researchers would ignite the thermite, thereby providing a few minutes of intense simultaneous temperature and pressure.

As soon as all the hardware was in place, ASEA's first order of business was simply to calibrate the beast. It was no easy task to determine the pressure and temperature at the center of such a device, and precise measurements of pressure and temperature were never obtained, although it now appears that pressures exceeding 60 000 atmospheres were generated in some experiments. The tedious and largely futile calibration process was followed by several abortive diamond-making runs. These experiments failed to produce much of anything except broken anvils and deformed copper jackets. ASEA workers had to labor as much as three to four months to assemble each experiment because so many of the components had to be replaced after each run. With no sign of success and years of seemingly wasted effort, ASEA considered abandoning the project.

Von Platen's apparatus – the reinforced tank, the massive yoke, the six mighty anvils, and all the other bits and pieces of exotic high-pressure machinery – were shipped back to the old Stockholm palace in 1948, and for a year or more very little was done. For a brief time the world's efforts to make diamonds seemed to have ceased altogether.

That might have been the end of von Platen's diamond-making dream, but for some reason the corporate powers-that-be changed their minds, and near the end of 1949 ASEA resolved to try again with twice the resources and a new team of five scientists and engineers. This time all the work took place at the Stockholm palace, where von Platen could be consulted on a regular basis, though the inventor played almost no direct role in the experiments. The top-secret diamond-making project was code-named QUINTUS, and the research facility was dubbed the Quintuslaboratorium.

* * *

ASEA epic experiments at von Platen's Stockholm laboratory took place in one of the most historic outhouses in the city. The giant yoke, weighing in at fourteen tons, was much too heavy for the floors of the main palace building, so the high-pressure laboratory had to be located in one of two elaborate bathrooms in the palace courtyard. The toilets were removed from one building, and it was outfitted with heavy-duty pumps, gauges, and other machinery. The adjacent outhouse retained its original function, and at least one summertime visitor recalled an unpleasant odor in the lab, perhaps unique in the annals of high-pressure science.

A senior ASEA research director, Halvard Liander, oversaw the project, while the younger, energetic Erik Lundblad took over the hands-on aspects of the work. Lundblad (fig. 22) is remembered as providing a striking contrast to the reserved von Platen. He was a big, bullish extrovert with an appetite for big meals, strong drink, good cigars, and (so the stories go) blondes. Erik Lundblad made friends easily and struck colleagues as more like a politician than a scientist. Some may have seen him as simply the front man for von Platen's genius, but his expertise and enthusiasm proved essential to ASEA's success.

It would be hard to overstate the drama surrounding each experiment of the QUINTUS project during those attempts of the early 1950s. The team first repeated the procedures used in earlier runs. Months of painstaking machining of new parts, dangerous shaping and packing of thermite, meticulous assembly to ensure proper alignment of all six anvils, and difficult copper jacketing preceded each run. Each of these complex steps had to be perfect or the entire effort would fail. Then the spherical high-pressure assembly had to be tightly sealed in the potentially explosive water tank, which could burst at any time from the 6000 atmospheres of internal pressure. Once the maximum pressure was obtained and stabilized, each experimenter must have held his breath, waiting to see if the deeply encased thermite would ignite properly. If that crucial step failed then all was for nought, and the whole aparatus had to be dismantled and assembled anew, with the added danger of extracting the unstable thermite, which could burst into sunlike flame at any instant.

Erik Lundblad recalled that the dismantling process "could be completed in one day unless everything exploded . . . This happened quite frequently and I can understand the feelings which Hannay so graphically described. . . . The situation became especially critical when we went over to [thermite-type] chemicals for heating. These were both explosive and toxic. Fortunately, no safety inspectors knew what we were up to."[6]

Fig. 22 Erik Lundblad tests electrical connections before an experiment with ASEA's diamond-making apparatus. (Courtesy of Erik Lundblad.)

Even if everything went exactly as hoped and high pressures and temperatures were achieved, there was still no guarantee of success. Many nervous hours must have passed as the water pressure was released, the tank removed, and the sphere disassembled. It would take days before all the run products could be carefully extracted and tested – dissolved in acid, examined by microscope, and x-rayed for the telltale signs of diamond. The months and years slipped away. Every ASEA run failed.

When one experiment doesn't work, the obvious strategy is to do something different, and the QUINTUS project scientists tried everything they could think of. In 1951 they focused on direct conversion of graphite to diamond. In some experiments the flat graphite crystals were carefully stacked, in others the orientation was varied. In 1952 they experimented with mixtures of carbon and metals in the sample assembly, but the sample appeared to react chemically with the thermite, and no diamonds were found. It was about this time that von Platen sold all rights to his high-pressure designs to ASEA and discontinued his consulting relationship with the firm.

Early in 1953 Lundblad and coworkers changed their strategy again and tried an iron carbide and graphite mixture that also contained a single small transparent diamond crystal as well as fine diamond powder. The reasoning behind this effort followed directly from the experiments of Moissan and others, who had showed that carbon dissolves in molten iron at high temperature and will crystallize out in its stable form when cooled at high pressure. Iron carbide provided the densest, most concentrated form of iron plus carbon available to Lundblad and his colleagues. They wrapped the central carbon-rich sample in a foil shell of platinum and tantalum to prevent a chemical reaction with thermite.

On February 16, 1953, the experiment was run under a pressure later estimated to be 83 000 atmospheres sustained for a full hour. Upon releasing the pressure, Lundblad and his coworkers disassembled the split sphere and recovered the sample; the foil-wrapped core was still intact. With exquisite care they unwrapped the foil, and quickly found the original single diamond crystal. It had emerged from the ordeal opaque, the result of surface cracks, but was otherwise intact – proof that the pressure and temperature achieved in the experiment was close to those at which diamond is the stable form of carbon. The scientists subjected the rest of the black sample to the closest scrutiny. The tedious process of repeated acid baths and microscopic analysis eventually revealed forty or fifty tiny crystals, the largest no bigger than a grain of sand. These crystals hadn't been in the starting materials, and subsequently x-ray analysis confirmed their hopes. The new material was diamond.

For millennia humans had held the gem in awe, and for centuries had sought for a way to make it. On February 16, 1953, that quest was ended.

The moment of discovery is remembered as one of unrestrained joy for ASEA's QUINTUS team, though it was a joy that could not be widely shared. The synthetic origin of the sample was still top secret – no one outside a small group of employees knew the source of the crystals. "The fellow who performed the x-ray analysis could not understand why I became so excited when he reported that my sample indicated only the presence of diamond," Lundblad recalled. "My God, I still get headaches thinking of how this was celebrated."

The historic success was repeated twice in 1953: on May 24th, when they synthesized diamond from a starting mix containing no seed diamonds, and again on November 25th, when the experimental products were independently identified by an expert not associated with the project. Each experiment is said to have yielded a miniature trove of tiny synthetic diamonds.

But ASEA's success was bittersweet. The unwieldy technique was difficult to reproduce and much too slow and costly to have commercial applications. While ASEA researchers struggled in secrecy to repeat their process, credit for achieving diamond synthesis went to a rival team of American diamond makers.

* * *

The most maddening, inexplicable aspect of ASEA's diamond-making victory was their absolute, absurd silence. After hundreds of years of concerted effort in which brilliant scientist after brilliant scientist from Hannay to Bridgman had failed, ASEA had triumphed. The first chapter of the diamond-making saga might have ended – should have ended – in 1953, if only ASEA and its researchers had followed normal scientific protocol. If they had announced and published their historic results, if they had filed a patent application, even sent a sealed document describing the breakthrough to a neutral party, their contribution would have been endorsed throughout the world. But they did none of these things. Instead, they waited and said nothing.

Two years later, after rival diamond makers in America had announced their own successful synthesis, ASEA made a brief and relatively uninformative mention of their diamond-making activities in the *ASEA Journal*.[7] A formal description of the synthesis procedures did not appear until 1960, seven years after the event.[8] Indeed, so much of the ASEA story was only told in retrospect – some of it decades after the original work – that many details of the experimental results, and even the motivations for the experiments, are difficult to document for sure.

Why ASEA hesitated to report the breakthrough has remained the greatest mystery. Part of the reason must have been their disappointment in making only diamond grit when magnificent gemstones had originally been envisioned. The cumbersome split-sphere process devised by von Platen could hardly have become economical unless each experiment produced a large, valuable diamond. But in a year of successful synthesis, employing five researchers and millions of dollars worth of hardware, ASEA had grown only a fraction of a carat of tiny diamonds. Their first successful experiments held little promise of commercial viability.

Erik Lundblad's explanation for ASEA's silence seems believable. "We at ASEA made no public announcement in 1953, mainly because we wanted to improve on the size and quality of the diamonds and also to gain more detailed knowledge of the conditions of diamond formation," he said. The thermite technique was dangerous, destructive, and far from being perfected. Any premature announcement would have just given rival companies a chance to catch up, and safety inspectors reason to shut them down. Furthermore, based on their conversations with Percy Bridgman in 1951, the ASEA group believed that no one else had joined the diamond-making game. "There we were apparently misled," Lundblad lamented.[9] Rather than publish their results right away, they tried to incorporate a more accurate electrical heater in the design and to adopt a simpler piston-in-cylinder-type mechanism for generating pressure. "We thought we had time," he said.[10]

A rather different account is advanced by publicists of the De Beers company, who joined with ASEA to form the Scandiamant diamond-making company in 1965. De Beers took full control of this firm in 1975, and since then has more or less assumed the role of partisan historian for ASEA's efforts. In a 1990 issue of *INDIAQUA*, a De Beers public relations magazine, the company presented what, at best, seems a revisionist view of the history.[11] "What, then, was the motive?" the article asks. "Hardly commercial, nor due to an actual need of the product. Despite the fact that diamond is rare, there was no actual shortage of diamonds. No, the interest as far as [corporations] were concerned lay in the challenge of being able to do something which no one else could do, and thereby gaining an even greater reputation for their respective companies." But if reputation was the key, and if ASEA had spent millions to make diamonds, confirmed the synthesis with x-ray diffraction, and then duplicated the feat in 1953, surely they would have announced their triumph immediately to gain the credit. Furthermore, in 1953, at the height of the Korean War, there *was* a diamond shortage, and a new potential source would have had tremendous significance. Why did ASEA wait for years after General Electric scooped them to make an announcement? The De Beers corporate explanation doesn't ring true.

Skeptics, particularly those associated with General Electric who admittedly have their own historical agenda, have proposed yet another version of events. Might the ASEA workers have failed to identify the tiny 1953 synthetic diamond crystals synthesized until *after* the 1955 GE announcement, they ask. Perhaps they were the first to make diamonds, but didn't realize it until two years later.

Whatever the reason for ASEA's silence – caution, secrecy, or ignorance – the delay of their announcement negated any claim they might have made for priority of discovery. Scientists have established certain rules for deciding who accomplishes something first. Most important, an experimental result must be fully described by the discoverer and then duplicated by independent researchers before it can be accepted. Full public disclosure is not required when proprietary procedures and equipment, like the ASEA diamond machine, are involved, but complete details of the process can still be written down, deposited in a sealed document, and placed in the custody of a neutral party. These actions must be taken prior to disclosure by a rival group.

Although they are not officially credited with the discovery, Erik Lundblad and his ASEA coworkers are often mentioned as possibly the first to synthesize diamonds. Baltzar von Platen's bizarre split-sphere device – expensive, ungainly, dangerous, and self-destructive – probably succeeded where all earlier attempts had failed. But ASEA's failure to publicize the feat until after General Electric's announcement has relegated the Swedish success to little more than a historical footnote.

Years after the event, von Platen regretted ASEA's failure to announce the creation of diamonds. "I write now as I should have written perhaps twenty years ago. Then, it was the high pressure and the great volume of the high pressure chamber which were prime considerations. They were the goal which had for so long slipped away like a mirage into the seemingly impossible. Yet when these things were finally achieved one found that there were new difficulties to be overcome."[12] By failing to share his secrets with the world, Baltzar von Platen lost his greatest opportunity for fame.

Notes

1. Baltzar von Platen, "A multiple piston, high pressure, high temperature apparatus," in R.H. Wentorf (editor), *Modern Very High Pressure Techniques.* (Washington: Butterworths, 1962), pp.118–136; quote on p.118.

2. Unpublished details and many quotations used in this chapter about the ASEA diamond synthesis effort were obtained in interviews and correspondence with Eric G. Lundblad. Additional information was provided by Francis Bundy, H. Tracy Hall, Herbert Strong, Alvin Van Valkenburg, and Hatten S. Yoder, Jr.

3. von Platen, op. cit., p.119.

4. Ibid., p.120.

5. The historic context and corporate policies related to ASEA's diamond-making efforts are reviewed in: Ekik Lundblad, "Swedish synthetic diamond scooped the world 37 years ago." *Indiaque* 55(1), 17–23 (1990).

6. Ibid., p.21.

7. Halvard Liander, *ASEA Journal* 28, 97 (1955).

8. Halvard Liander and Erik Lundblad, "Some observations on the synthesis of diamonds." *Arkiv für Kemi* 16, 139–149 (1960).

9. Erik Lundblad, "Letter from Dr. Erik G. Lundblad." *Journal of Gemmology* 20, 137–138 (1986).

10. Gordon Davies, *Diamond.* (Bristol: Adam Hilger, 1984), p.88.

11. See, for example, Kurt Nassau, "A note on the history of diamond synthesis." *Journal of Gemmology* 19, 660–663 (1985). See also letter in reply: Erik Lundblad, ibid. 20, 137–138 (1986).

12. Von Platen, op. cit., p.135.

CHAPTER 5

THE CRYSTALS OF LORING COES

Using new techniques, it was found possible to synthesize most of the well-known high-pressure minerals.... The main motive in this work was to study the conditions attendant on the formation of natural diamond.[1]

LORING COES, "HIGH-PRESSURE MINERALS," 1955

B Y THE LATE 1940S AMERICAN efforts to synthesize diamonds had acquired a new urgency as Cold War tensions gripped the nation. Only diamond could machine and polish the carbide tools needed to cut and shape the critical components of aircraft engines, gun barrels, vehicle armor, and other military hardware. Except for an emergency stockpile of industrial diamond in Canada, North America was completely dependent on South Africa, a vulnerable target easily cut off from the West, to satisfy the growing demand for diamonds. The nation turned its hopes to the diamond makers.

Percy Bridgman had taken up the synthetic diamond challenge in 1941 at the behest of the Norton—GE—Carborundum consortium, but he abandoned the effort for other research during World War II and was disinclined to return to the problem afterward. Bridgman must have realized that his experimental skills did not extend to the high-temperature technology necessary for diamond making. Rather than give up on high-pressure research altogether, the Norton Company decided to continue the project on its own. They moved Bridgman's thousand-ton press to their Worcester, Massachusetts, research laboratory and began to think about new ways to make diamond.

From 1945 to 1955 Norton employees engaged in a variety of high-pressure experiments, but the work of chemist Loring Coes, Jr., yielded the most important discoveries. Bridgman's work had almost always relied on pressure alone, and he spent little effort designing furnaces or electric heaters, but Coes took a different approach. Taking his cue from how diamonds formed in nature, Coes started with rather ordinary mixtures of chemicals, heated them to more than a thousand degrees, squeezed them to tens of thousands of atmospheres, and, as often as not, came up with something no one had ever made before.

* * *

Loring Coes, Jr., the second of five children and the oldest of three boys, was born in Worcester, Massachusetts in 1915.[2] He grew up in nearby rural Brookfield, a farming community where he developed a lifelong passion for riding horses and chemistry. Even as a high school student Coes was fascinated by chemical reactions and he built a laboratory in his home. Small in stature and unassuming by disposition, he was remembered by his friends as a taciturn man, difficult to get to know. He was inclined to smoke too much and drink too much, but proved to be an intense and creative scientific researcher. After graduating with bachelor's and master's degrees in chemistry from Worcester Polytechnic Institute in Massachusetts, he became a researcher for the Worcester-based Norton Company, a high-tech firm with a strong interest in superhard materials. Coes's triumphs in high-pressure research represented a high point in a life ultimately marked by personal failures and disappointments.

Coes was a meticulous, dedicated scientist who worked long hours in his Worcester lab and at home. For four decades he remained a loyal employee at Norton, where he impressed colleagues with his brilliant innovative thinking and unusual technical skills. An outstanding chemist, Coes brought an intimate knowledge of chemical reactions, synthesis techniques, and properties of materials to his research. An accomplished glass blower, he fabricated all of his own experimental glassware as well as glass vessels for the Norton analytical laboratory. He also had an extraordinary memory and a gift as a precise and articulate public speaker.

Coes's career at Norton encompassed many research projects, from work on organically bonded ceramics to the development of the theory of grinding, but he is remembered by scientists today almost entirely for Project A-39, a brilliant six-year foray into the high-pressure synthesis of minerals. This diamond-making project was the brainchild of Samuel Kistler, Norton's imposing, inspiring Associate Director of Research. Standing six-foot-four with thick-lensed glasses and a professorial air, Kistler was a walking encyclopedia, a lively personality with an abundance of ideas and enthusiasm – a perfect foil for Milton Beecher, Norton's conservative Vice President for Research. It was Kistler who pointed Loring Coes toward high-pressure science.

Coes was joined by a small team of coworkers. Max Wieldon, a gregarious and athletic forty-year-old mechanical engineer with a degree from Middlebury College, focused on designing and building high-pressure apparatus. George Comstock, recently graduated from Worcester Tech, searched for the strongest pressure-vessel materials and studied ways to calibrate pressure and temperature. Chinese-born research assistant Wing Moy assisted Coes in the laboratory, while machinist Peter Jensen, a Danish immigrant tool and die-maker, was also critical to the effort.

Coes was well aware of the earlier attempts to make diamonds – except, perhaps, for the secret ASEA project – and he knew about the difficulty of generating sufficient simultaneous high temperature and pressure. So rather than tackle diamond synthesis immediately, he decided to creep up on the conditions necessary for the gem's formation by making some of the other forty or so known dense, deep-earth minerals that had never been produced in the laboratory. Not only was this research of great scientific interest, but it also seemed to Coes to provide the most promising avenue for finding tough new abrasives – and for uncovering the secrets of diamond synthesis.

Percy Bridgman's work had provided one clear message to would-be makers of diamond and other deep-earth minerals: pressure alone is not enough. You must heat while you squeeze. Unfortunately, high temperature weakens steel or carbide. Coes had to devise a mechanism strong enough to support unprecedented pressures at extreme temperature. Most pressure devices failed because all the force was directed along just two opposed pistons. Failure occurred when the pressure first deformed the metal ring surrounding the sample, which then blew out sideways, like a grape bursting in a wine press. The key was to find a way to confine all sides of a sample at once. Von Platen and his ASEA followers had gotten around the problem by using six anvils, each pressing on a side of a cube, but that device was expensive and ungainly, and ASEA's methods were top secret to boot. Coes's team had to find their own way.

Most high-pressure scientists with access to a huge press like the Norton thousand-ton machine would have been seduced into working on a large scale with bulky devices and big samples, but Coes and his colleagues thought differently. Pressure is force per area: you can create a lot of pressure by subjecting a large surface to a huge force, but you can do it much more easily by subjecting a very small area to a modest force. Furthermore, as a chemist, Coes knew that sustained high temperatures are often a lot easier to achieve with smaller samples. So the Norton team got rid of the big press (it was modified for routine manufacturing chores) and they turned their attention to building a smaller, beautifully crafted piston-in-cylinder machine.

Coes, Kistler, and Wieldon built their simple device from two high-tech materials. The first material, tungsten carbide, was well-known to high-pressure workers thanks to Percy Bridgman's experiments, but the ultratough compound was not generally available to researchers. Only a handful of U.S. companies controlled its distribution; fortunately, Norton was one of them. Two tungsten carbide pistons, each one-quarter inch in diameter, applied force on the sample from either end of a cylindrical steel sample chamber with a quarter-inch central hole.

To make the sample cylinder, another extraordinary high-tech material was used – a hard and rigid ceramic dubbed Alundum, available as a specialty product from Norton. Ceramics are wonderfully useful materials. They make up substances like china, bricks, and most rocks, all of which are composed primarily of microscopic interlocking crystals. Ceramics can be found almost everywhere, from coffee cups and bathroom tiles to high-voltage electronics and the space shuttle's exterior insulation. They are typically very hard, usually act as good electrical insulators, can resist extremely high temperatures, and have a tendency to shatter when dropped.

Norton had developed a variety of ceramics under the trade name Alundum, a reference to the material's principal constituent, aluminum oxide or, in its natural form, the hard mineral corundum. The gemstones ruby and sapphire are nothing more than exceptionally rare colored forms of this common, normally colorless compound. Coes might have contemplated using cylinders of single-crystal corundum in his press, but large single crystals are rare in nature, difficult to grow in the laboratory, almost impossible to machine, and they tend to break along the lines and planes of crystal weakness. Single crystals of corundum, even if available, would make poor high-pressure components.

Alundum, on the other hand, could be easily made by compressing and heating powdered corundum in a precisely machined mold. This process, known as hot-pressing, causes the tiny powdered grains to recrystallize, cementing together to form a polycrystalline mass with virtually no pores or directions of weakness. Coes selected an Alundum recipe using the purest aluminum oxide available and had the Norton technicians make him a steady supply of perfect cylinders with an inner diameter exactly one-quarter inch to match the tungsten carbide pistons. Alundum was the perfect choice for the job.

The Norton workers pressed the ceramic cylinder into a thick, tightly fitting, outer ring of steel with slightly tapered hole. This metal binding ring reinforced the rigid ceramic cylinder, preventing the sample from bursting through the cylinder walls. A larger cylindrical water-cooling jacket fit around the entire piston-cylinder mechanism to complete the simple assembly. Coes's press thus took full advantage of the materials expertise of the Norton Company – one of the few companies in the world that could have supported his high-pressure effort.

Perhaps the most brilliant and elegant feature of Coes's device was its heating capability. Aluminum oxide, which formed the cylindrical walls of the sample chamber, is an excellent electrical insulator, while the tungsten carbide pistons that formed the chamber's top and bottom surfaces are good electrical conductors. To produce extreme tempera-tures, all Coes needed was an electrical heater between the two carbide pistons. For this task he turned to an old standby: graphite. By attaching electrical leads to the two carbide pistons, bridging the gap between the pistons with a graphite cylinder, and turning on the juice, Coes could heat samples to more than 1000°C by using graphite's electrical resis-tance. Furthermore, because a graphite heater was present at the heart of every experiment, no matter what the mineral under study, Coes had an excellent opportunity to make diamond as well. Simple fifty- and hundred-ton hydraulic presses, a half dozen in all, were all that Coes and his team needed to squeeze a steady stream of samples.

At the extreme pressures and temperatures of Coes's experiments, almost every run resulted in the breakage of the aluminum oxide cylin-der. The tungsten carbide rods, which were hard but brittle, and deformable graphite cylinder heaters also needed frequent replace-ment. Yet unlike the cumbersome ASEA device, which took months to repair after each experimental run, Coess' simple system took only an hour or so to make ready for the next experiment. None of the parts was large or difficult to make, and Norton technicians kept Coes supplied with all the pistons, cylinders, and graphite heaters he needed to keep the operation running efficiently.

The research agenda was almost as simple as home cooking. Measure out common chemicals like magnesium oxide, silicon carbide, and aluminum hydroxide. Thoroughly mix and grind the powders and enclose the mix in a metal foil capsule, usually of copper or iron. Heat and squeeze for an hour or two. Open the capsule and, voilà, look at the beautiful crystals.

Although this procedure may sound relatively simple, Coes had a few secrets to ensure success. All of the dense minerals that he attempted to make were silicates – compounds of silicon and oxygen, usually with one or two other elements. The most logical synthesis approach would be to mix these elements together as powders, and the simplest powders to use were oxides – mixtures of oxygen and a second element. If you wanted to make the high-pressure garnet pyrope – a silicate with magnesium, aluminum, silicon, and oxygen, for example – the normal, reasonable procedure would be to weigh and mix carefully the correct proportions of magnesium oxide, aluminum oxide, and silicon oxide; seal these oxides tightly in a capsule so nothing could escape, and heat and squeeze. Unfortunately, this procedure won't make pyrope.

Coes defied conventional logic. To make pyrope he mixed magnesium nitrate, aluminum hydroxide, and silicon carbide, a chemical mix many earth scientists found absurd. Then, instead of sealing his starting materials tightly inside metal capsules, he simply crimped the ends closed, purposely allowing impurities to "spoil" his experiments. And Coes made beautiful crystals of pyrope, nearly every time he tried.

What at first appeared to be sloppy procedure was in fact good chemistry. It turned out that Coes's "secret" ingredients made the atoms in his samples much more mobile and reactive than in simple oxides. The slightly open capsule allowed all the extra volatile material in Coes's mix – water, nitrogen, carbon dioxide, and hydrogen – to escape, leaving behind perfect crystals of garnet. Time after time he grew silicates from unlikely starting materials – nitrates, sulfates, carbides, even chlorates. His crystals were tiny but exquisite, with perfect geometric faces, and his technique was so simple that a single worker could complete three or four experiments in a single day.

Unlike the ASEA team, Coes hit pay dirt almost immediately. When Coes applied high temperature and pressure to his novel mixes of chemicals, he produced an extraordinary range of mineral-like phases that normally only occur miles deep within the earth. In his first experiments he focused on the minerals found in eclogite, a beautiful deep-earth rock, occasionally diamond-bearing, that displays deep red garnets and stunning green pyroxenes. Garnets, pyroxenes, and other eclogite minerals grew in abundance in his laboratory. Coes also synthesized several common metamorphic minerals – crystals like staurolite and kyanite that form when ordinary mud, silt, and other sediments are buried and baked many miles deep within the earth. Geologists had long known about these minerals from metamorphic rocks in the Appalachians, the Alps, the Urals, and other ancient mountain ranges, wherever thick piles of cooked sediments have been brought back to the surface. Loring Coes was the first to synthesize these minerals in the laboratory.

Semiprecious gemstone minerals formed with ease in Coes's laboratory. In his press he grew crystals of zircon, idocrase, tourmaline, beryl, sphene, and topaz – dozens of different minerals were created in hundreds of runs. Coes used an old brass petrographic microscope, which had special optical attachments favored by geologists, to identify and describe his growing cache of experimental products.

Coes's most stunning discovery, by far the biggest bombshell occurred when he squeezed one of the simplest, most common minerals of all, ordinary beach sand, known as quartz – a form of silicon oxide, or silica. Before this work on quartz, his experiments had always produced well-known minerals. But when Coes squeezed quartz to 35 000 atmospheres at 800°C he found a new dense silica phase, a crystal substance never before seen. This unexpected find had two profound implications. First, the new form of silica was dense and hard, perhaps not a commercially viable abrasive, but better by far than most known minerals. Coes thus proved what many had suspected – that previously unknown superhard materials might be found at high pressure. More importantly, however, Coes had produced astonishing direct evidence that less than fifty miles below the earth's surface there might be other dense minerals that had never been seen before. If a mineral as common as quartz could occur in an alternate form, what other strange and wonderful minerals might exist beneath our feet?

* * *

From 1947 until 1953 Loring Coes worked in obscurity, his extraordinary results were completely unknown to the outside world. Because he served an industrial laboratory where discoveries counted as corporate assets, sharing those discoveries was seldom a high priority. Furthermore, Coes was by nature a quiet and introspective man who hesitated to trumpet his own accomplishments. But upon discovering the transformation of quartz, he was given permission to contribute a brief technical note, "A New Dense Crystalline Silica," which appeared in the July 31, 1953, issue of the prestigious weekly journal *Science*.[3] That short abstract – a condensation of several years' work – sent a shock wave through the earth-science community. "The possibility exists," he wrote, "that the existence of this form of silica in nature may have been overlooked." In form, color, and optical behavior the new dense silica could easily be mistaken for other common minerals, he suggested. Coes, an author completely unknown to his audience, tantalized readers with the promise that "A subsequent paper on the synthesis of several naturally occurring minerals will greatly amplify this information."

The mineralogy of the earth's interior – almost all of the solid earth's volume – was literally terra incognita in the 1950s, and the geological community eagerly sought any new technique that might open up the earth to laboratory investigation. Within a few weeks of the Science (fig. 23) report, Hatten S. Yoder, Jr., a dynamic young high-pressure

Fig. 23 Loring Coes (left) with Hatten S. Yoder, Jr., in Yoder's high-pressure laboratory, January 31, 1978, shortly before Coes's death. (Courtesy of Hatten S. Yoder, Jr.)

researcher at the Carnegie Institution of Washington's Geophysical Laboratory, invited Coes to come to the nation's capital. Yoder had been trying for years to make high pressure minerals. He had accomplished much in a device that used extremely-high gas pressure, but synthesis of minerals like pyrope garnet and kyanite from pure oxides had eluded him. He received the news of Coes's success with a combination of excitement and skepticism.

Coes made the trip to Washington on Monday, September 21, 1953, and provided Hat Yoder with a brief outline of what he had accomplished in the Norton lab. Much of the research was proprietary, and many details could not be discussed at that time, but what Coes did reveal – the new minerals synthesized in his lab and the new materials used to construct his high-pressure and high-temperature device – was more than enough to whet the appetite of the Lab's high-pressure scientists. Yoder knew he had to visit Worcester to see for himself.

Following Norton's official okay, Yoder contacted Harvard geophysicist Francis Birch, who assembled a contingent from Cambridge, Massachusetts, to meet Yoder at Norton. He also invited Alvin Van Valkenburg of the Washington-based National Bureau of Standards to join the expedition.

The memorable Friday afternoon meeting took place at 2.00 pm on December 4th, 1953 (fig. 24).[4] The Washington contingent included Van Valkenburg, Yoder, and Francis "Joe" Boyd. Francis Birch, post-doctoral fellows Eugene Robertson, Gordon MacDonald and Jim Thompson, and the distinguished mineralogist Cornelius Hurlbut rounded out the Harvard group. Birch and Robertson were particularly intrigued by Coes's revelations because they had been struggling for almost two years on a contract with the Office of Naval Research (ONR) to study "Properties of Materials at High Pressures and Temperatures," a project whose unspoken, ultimate goal was to make diamonds.

The scientists from Harvard and Washington were also joined that day by Rustum Roy, a high-pressure expert from Penn State and a remarkable figure in high-pressure history. Roy graduated with a Bachelor's degree from India's Patne University, and in 1946 arrived at Penn State as a graduate student interested in micas, one of India's major mineral exports. As he was finishing his Ph.D. thesis in 1948, Roy was asked to stay on as a postdoctoral fellow by E. F. (Ozzy) Osborn, who had just received an ONR contract for "Experimental Metamorphic

Fig. 24 The December 4, 1953, meeting at the Norton Company in Worcester, Massachusetts, brought together many leaders in high-pressure research. At the meeting Loring Coes demonstrated his novel method for synthesizing high-pressure minerals. Seated, from left to right: Loring Coes, Jr., Cornelius Hurlbut, Francis Birch, Hatten S. Yoder, Jr., Alvin Van Valkenburg. Standing: Rustum Roy, Francis R. Boyd, James Thompson, Gordon MacDonald, Eugene Robertson. (Courtesy of H. S. Yoder, Jr.)

Petrology" – in other words, money to make high-pressure rocks in the lab.

In a short, immensely productive period in late 1948 Osborn and Roy designed apparatus and techniques for what they called routine hydrothermal synthesis of samples at pressures up to a few thousand atmospheres. They encapsulated water-bearing mixes of elements in gold and transformed the contents with pressure and heat. By late 1953 the Penn State group had studied numerous mineral systems, but repeatedly failed to make the high-pressure metamorphic mineral kyanite. Roy was alerted to the special December 4th meeting by another Norton employee, and he had to be there to see for himself.

It was an extraordinary group: Birch, Boyd, Hurlbut, MacDonald, Robertson, Roy, Thompson, Van Valkenburg, and Yoder. Coes showed amazing things to this vertiable who's-who of American high-pressure scientists.

"We arrived as a bunch of doubting Thomases" remembers Al Van

Valkenburg, whose high-pressure work at the National Bureau of Standards focused on the optical behavior of minerals. "No one believed he could have made the things he claimed."

"We were all skeptics," Rustum Roy echoes. Some scientists believed that the formation of deep-earth minerals depended on complex stress patterns and control of rocks' water content – factors not easily reproduced in the laboratory. They felt that the only rocks you could hope to mimic in the laboratory were those found in near-surface deposits. After all, a lot of researchers had tried and failed in the quest for deep-earth minerals.

It didn't take Coes long to change everyone's mind. When they handled the sophisticated tungsten carbide and hot-pressed Alundum equipment, when they saw the crystal specimens – many large enough to pick up with tweezers – when they looked in the microscope and read the unambiguous x-ray diffraction patterns that identified each high-pressure mineral, there could be no doubt. Gene Robertson remembers the meeting as an extraordinary learning experience. "It was clear right from the start he knew what he was doing." Not only had he created numerous synthetic minerals in novel ways, with combinations of starting materials that no one had thought of before, but he accomplished this feat in an elegant apparatus made from materials that the visiting researchers had never seen before.

"He made all the really high-pressure [minerals], which were like Holy Grails at the time," recalls Roy. Norton Company policy prohibited the visitors from taking photographs or writing notes, so the eager scientists had to rely on their memories to retain the information that Coes offered. Hat Yoder remembers visiting a mens' room stall at one juncture and furiously writing everything he could recall on the back of the only useable scrap of paper at his disposal, a blank check in his wallet. That check, still preserved in his notes, is crammed with tiny formulae of strange starting materials and extreme conditions of synthesis.

Coes later bent the rules a bit, giving Yoder a Polaroid black-and-white photograph of the neatly arranged pieces of the device. On this photo Coes labeled each piece in crisp printing – "hot pressed alumina mold liner," "WC [tungsten carbide] pistons," "steel mold retainer," and all the rest.

"It was a powerful stimulant," Joe Boyd recalls of the visit. "It really shook people up." Overnight, Coes' pioneering syntheses changed the way scientists thought about investigating the earth. The nine scientists left Norton amazed; none of them would ever see the world in quite the same way again.

* * *

Upon his return to the Geophysical Laboratory, Hat Yoder wrote to Coes with the thanks of all the participants. "We all greatly appreciate the fine tour of your laboratory and the opportunity to see some of the details of your apparatus. You probably do not realize what a tremendous contribution you have made to the field of mineral synthesis and to geology in general. The impetus you have given our work is very great."[5]

Yoder's words were no exaggeration. In one way or another, the December 4th meeting profoundly influenced all the participants. Hat Yoder began to construct a copy of the Norton device almost immediately upon returning to the Geophysical Laboratory. He lacked access to the sophisticated tungsten carbide and hot-pressed alumina components, but he substituted the highest grade steels available. For the dramatic first run he invited the Lab's newly appointed Director, Philip Abelson, to watch. The experiment had barely gotten underway when a sickening crack was heard. The pistons had broken and had become hopelessly, permanently jammed in the cylinder. Tungsten carbide pistons and hot-pressed alumina cylinders, items not yet commercially available from Norton, were evidently critical to success. Yoder went back to work (and to make history) with his own invention, a 10 000–atmosphere, gas-pressure apparatus that could accurately reproduce the full range of temperature and pressure conditions found in the earth's crust.

Joe Boyd, then a Geophysical Lab postdoc fresh out of graduate school at Harvard, had already resolved to study minerals in the new high-pressure, high-temperature regime typical of Coes's work. After the historic meeting he designed and built a piston-cylinder device of his own and proceeded to study the behavior of silicate minerals at deep-earth conditions. One of his first studies at the Geophysical Lab, made in collaboration with long-time colleague Joe England, was the careful measurement of the exact pressures and temperatures at which quartz transforms to the new form of silica discovered by Coes.[6]

The Norton meeting was also pivotal for Harvard Professor Francis Birch and his postdoctoral fellow, Gene Robertson.[7] Perhaps more than anyone else, Birch had carried on Percy Bridgman's legacy at Harvard.

After studying with Bridgman as a Harvard physics graduate student in the early 1930s, and spending World War II at Los Alamos working on the Manhattan Project, he continued pressure research at Harvard's Department of Geological Sciences. Birch and Robertson had labored to produce diamond for ONR, with little success. The project, begun in 1952, relied on a Bridgman-type device to attain sufficient pressure, but the researchers lacked the tough new anvil materials necessary to make significant progress. Coes graciously offered to help by supplying hot-pressed alumina components. Birch and Robertson never did make diamonds as originally hoped, but the Norton expedition did point them in promising directions, and they went on to make critical measurements on jadeite, kyanite, and pyrope – all minerals first synthesized by Coes.

Rustum Roy and his Penn State colleagues were also inspired to try something new. In collaboration with colleague O. F. Tuttle, Roy spent a year or so engaged in what he describes as an "amateurish look" for diamonds. The project was underwritten by the Carborundum Company, but they were never able to achieve sufficient pressures. In subsequent years, however, Roy and his associates used a Bridgman-style opposed-anvil device that could generate pressures up to 100 000 atmospheres at 550°C to produce more than a hundred new compounds, including novel forms of phosphates, fluorides, oxides, and metals.[8] Roy credits Coes for first pointing him on his way.

The December 1953 visit had another consequence as well. The revelations that came from that experience, coupled with the widely lauded discovery of a new dense form of silica, practically made Loring Coes a living legend in the earth science community. Less than a year after Coes published his discovery of the synthetic silica, Robert Sosman, an earth scientist at Rutgers, wrote a letter to *Science* advocating an appropriate name for the new compound.[9] "Fearing that the discoverer might be too modest to name the phase after himself," Sosman proposed the name "coesite" for the high-pressure form. Sosman apologized to mineralogical purists, who might object to giving an official mineral name to a synthetic material. "As an alternative for the benefit of any reader who wishes to stand firmly on the mineralogist's principle, I suggest that he call Coes' new phase of silica *silica C*." Everyone called it coesite.

Inevitably, just a few years after Coes's announcement of the synthetic silica, the same substance was found in nature. Edward Chao, a geologist with the United States Geological Survey, was studying effects of shocks on quartz by examining sandstone blasted by the Canyon Diablo Meteor at Meteor Crater, Arizona. Scrutinizing the material under his microscope, Chao noticed tiny crystals quite distinct from quartz. On a single day, an exhilarating Monday early in 1960, Chao isolated the material and identified it as Coes's silica compound.[10] The extreme shock of the meteor's impact had converted everyday quartz to its dense high-pressure form.

For a short time Chao wanted to exercise his right as discoverer to name the mineral "boydite," in recognition of Joe Boyd's ongoing research in the high-pressure mineralogy of silica, but the unauthorized name "coesite" had gained a strong, and deserved, foothold. The International Commission on New Minerals, ultimate arbiter of such matters, officially approved the name coesite in 1960, in honor of the man who first synthesized it.

* * *

Loring Coes gave scientists new hope in their quest for diamond. If so many deep-earth phases could be reproduced in the laboratory, why not diamond? But Coes never became a central figure in that adventure. In 1952 he was appointed Norton's Assistant Director of Research and Development, a job that diluted his effectiveness as a researcher. Alan G. King, a researcher under Coes at that time, remembers Coes's unique directoral style: he left his people alone. "I've never been under so much pressure in my life," King recalls. "You really try to excel when you're thrown into something like that."

The year 1962 was tragically pivotal for Loring Coes. As the scientifically better qualified of two candidates for the prestigious Director of Research at Norton, he looked forward to moving up the corporate ladder. But management found Coes's introspective manner and scientific inclinations incompatible with the top management job. On more than one occasion he had bluntly refused to pursue what he perceived as impractical research directions, in opposition to his superiors' wishes. The Norton CEO passed him over, a blow from which he never fully recovered.

In November, 1962 Coes was given a new and ostensibly prestigious research position, "Consultant in Research and Development," but it was little more than a consolation prize. Depressed and subject to bouts of alcoholism, Coes saw his life begin to fall apart. He was arrested for

Fig. 25 High-pressure pioneers Alvin Van Valkenburg, Loring Coes, Hatten S. Yoder, Jr., and F. R. (Joe) Boyd (left to right) at the Geophysical Laboratory of the Carnegie Institution of Washington, 1978. (Courtesy of H. S. Yoder, Jr.)

drunk driving and lost his license. His marriage failed, and his wife took all the furnishings, leaving him with little more than a cot in his large house. Though he remained a researcher with Norton for another sixteen years, he never repeated his extraordinary research success of the early 1950s (fig. 25).

Lung cancer, the consequence of a lifetime of heavy smoking, killed Loring Coes in 1978 at age sixty-three. He died in relative obscurity, with little more than a local obituary to mark the passing of the man who had transformed the international high-pressure scene.

* * *

Throughout the half-dozen years of Coes's high-pressure research, and in scientific publications years later, the search for high-pressure minerals was always cited as the principal scientific motivation for the Norton research. But Norton's research head, Sam Kistler, had actually begun Project A-39 in the hope of making diamonds, and almost from the start rumors circulated that traces of the precious material had been recovered from experimental runs. It appears that these samples were never subjected to rigorous testing, and it seems doubtful given the relatively low pressures involved that diamonds were ever made, though Coes fervently believed that in several experiments with diamond seed crystals he had succeeded in adding a thin layer of diamond to the original stones.

Although Coes abandoned his high-pressure synthesis studies in 1953, his assistant, Paul Keat, took over the project and for two years focused his efforts on creating diamonds. Realizing that higher pressures were essential for success, Keat obtained a larger press and concentrated on building a bigger device with opposed, tapered tungsten carbide anvils and a strong girdle to confine the sample. The project was well conceived and, had there been no competition, it stood every chance of success. But Norton's effort proved to be too little and too late; others had gotten there first.

In 1950, after three years of high-pressure studies, Coes and the Norton management had come to the conclusion that they were not going to make diamond, at least not with the double-piston device. They had enjoyed tremendous success creating high-pressure minerals, but diamond making would clearly take something more. In what may have been their most significant contribution to the history of diamond making, the Norton Company approached General Electric about a possible joint effort. Norton hinted that they had a "sniff of diamond" in some of Coes's synthetic runs and described the unprecedented synthetic products they had already created in their lab. On that basis, without revealing any information about the key experiments or the techniques employed, Norton asked GE to become their collaborator.

It must have been a tempting offer, and General Electric officials and lawyers tried to draft a cooperative arrangement. But a satisfactory agreement was slow in coming, and after considerable thought, the General Electric management declined Norton's proposal. Instead they decided to do it themselves.

Notes

1. Loring Coes, Jr., "Synthesis of minerals at high pressures." In: Robert H. Wentorf (editor), *Modern Very High Pressure Techniques*. (Washington: Butterworths, 1962), pp. 137–150.
2. Many details about Loring Coes, Jr., as well as his high-pressure synthesis

program at Norton were obtained through interviews with his friends and colleagues, including: Neil Ault, Samuel Coes, George Comstock, Paul Keat, Alan G. King, Rustum Roy, Lou Trostle, and Osgood Whittemore. Many technical details regarding these experiments are published in: Loring Coes, Jr., "High pressure minerals." *Journal of the American Ceramic Association* **38**, 298 (1955).

3. Loring Coes, Jr., "A new dense crystalline silica." *Science* **118**, 131–132 (1953).

4. Details regarding the meeting at Norton have been provided through interviews with: Francis Boyd, Gene Robertson, Rustum Roy, Alvin Van Valkenburg, and Hatten S. Yoder, Jr.

5. H. S. Yoder, Jr., to L. Coes, Jr., December 8, 1953.

6. Francis R. Boyd and Joseph L. England, "The quartz-coesite transition." *Journal of Geophysical Research* **65**, 749–756 (1960). An earlier effort, also inspired by Coes's work, is: Gordon J. F. MacDonald, "Quartz-coesite stability relations at high temperatures and pressures." *American Journal of Science* **254**, 713–721 (1956).

7. Francis Birch, "Reminiscences and digressions." *Ann. Rev. Earth Planet. Sci.* **7**, 1–9 (1979).

8. Frank Dachille and Rustum Roy, "High-pressure region of the silica isotypes." *Zeit. Kristallographie* **111**, 451–461 (1959).

9. Robert Sosman, "New high-pressure phases of silica." *Science* **119**, 738–739 (1954).

10. E. C. T. Chao, E. M. Shoemaker, and B. M. Madsen, "First natural occurrence of coesite." *Science* **132**, 220–222 (1960). I thank Edward Chao for providing unpublished details regarding the discovery of natural coesite.

CHAPTER 6

PROJECT SUPERPRESSURE

The history of attempts at diamond synthesis probably started in 1797, almost at the moment diamond was first shown to be a form of carbon.... The decision to mount a major new assault on this old and bloody battle-field was made in 1951.[1]

C. GUY SUITS, "THE SYNTHESIS OF DIAMONDS," 1960.

B ABY BOOMERS REMEMBER THE 1950S as a time of unbridled, can-do optimism. Americans had built the bomb, and had won the War. We were the richest, strongest nation on earth; there was nothing we could not do. It was a time when almost everyone shared in the benefits of wonderful new technologies – jet planes, sleek cars, and colorful plastic. People's lives were transformed by television, the pill, and credit cards. Medical miracles from the polio vaccine to open heart surgery raised life expectancy above sixty-five for the first time in human history. Amid such prosperity and confidence the time was right to make diamonds.

In earlier days the problem of diamond synthesis was tackled by individuals. Clever, dedicated scientists such as Hannay and Moissan had taken on the challenge like solitary knights on a quest. Time after time, these mavericks failed.

By the post-war years, high-tech research had changed forever. The many awesome new technologies spawned by war – developments like radar, jet aircraft and, of course, the atom bomb – were the products of large group efforts. World War II hammered home the lesson that teamwork was the answer to achieving big goals – a concept not lost on the post-war diamond makers.

No one person could pull off the diamond making trick, for nobody could master all the necessary expertise. Steel devices didn't work, so you had to have a materials expert with access to the best quality carbide components, as well as machinists skilled enough to craft that carbide into anvils or pistons of precise dimensions. Chemical heating with thermite was dangerously impractical, so you needed an expert electrician with the knowledge to heat a sample by electric current and the ability to do it in a cumbersome high-pressure system. Direct conversion of graphite to diamond didn't work, so you had to have a master chemist to discover the right recipe of chemicals to squeeze.

And trying to make diamonds was only part of the battle; you also had to isolate and identify the experimental products, a process that was complicated by the minute size of the crystals. Many researchers had been fooled by tiny diamond-like grains of other hard materials – spinel, corundum, chromite, silica, and carbide were all found to grow in a high-pressure environment and all formed hard crystals with triangular facets, just like diamonds. So you had to have an expert in analytical chemistry and x-ray crystallography just to prove that you made the real stuff. And, finally, you had to have deep pockets and a lot of time.

Diamond synthesis required so many different specialists – individual contributions from physicists, chemists, engineers, and money men – that bitter controversies over proper credit were perhaps inevitable. Decades of contentious debate between volatile personalities, coupled with a well-oiled corporate publicity campaign, have distorted and confused this dramatic history. What is clear is that by 1950 all the easy approaches to diamond making had been tried and had failed. Without the financial backing and long-term commitment of a determined corporation, the chances of making diamonds were slim.

* * *

General Electric Company was born a century ago, the brainchild of Thomas Edison, whose revolutionary innovations in electrical power, lighting, and transportation transformed the world. From the earliest days, GE corporate research and development carried on the Edison tradition, securing the company's fame and fortune. Invention after invention poured from the labs. They pioneered improved lighting with long-life tungsten filaments, soft-white frosted glass, and fluorescent bulbs. They transformed domestic life with countless new appliances, from refrigerators and air conditioners to toasters and electric blankets. They introduced mobile radio systems, modern-style x-ray tubes, and advanced propulsion systems for jet planes, electric trains, and ocean-going ships. By the first decades of the twentieth century, Edison's company had changed the way Americans used energy, and the GE logo was recognized world wide as a symbol of American excellence and know how.

The diamond game was serious business to GE, which depended entirely on expensive supplies of foreign diamonds to cut and shape their carbide products and to draw out the fine tungsten filament wire required for light bulbs. For that reason, Zay Jeffries, manager of GE's wholly owned subsidiary Carboloy Company, first approached Percy Bridgman about making diamonds by substituting carbide anvils for his weaker steel ones. General Electric assumed a leading role in the 1941 consortium that commissioned Bridgman's diamond studies, and in 1950, with no significant progress from other companies' research efforts, GE committed its own considerable resources to diamond making.[2]

The diamond-making effort, code named "Project Superpressure," was undertaken at the General Electric Research Laboratory in Schenectady, New York, a bustling city of 100 000. Located on the scenic Mohawk River a dozen miles northwest of Albany, Schenectady is a town of industry, with sprawling factories and vast rail yards. It is a city with a clear purpose – a city that makes useful products and ships them off for Americans to buy. It is not the first place you'd think of when talking of diamonds, but it was a place that knew how to get things done.

General Electric's first step was to find an experienced project manager, to keep the team running smoothly. They turned to Anthony J. Nerad, a man remembered with tremendous respect and affection by his former colleagues. His was an American success story. His parents emigrated from Czechoslovakia and settled in Milwaukee, and Tony took advantage of his opportunities by earning a degree in engineering from the University of Wisconsin. He gained renown at General Electric for a number of innovations, most notably his development of a critical jet engine combustion chamber that is still in use today. He was also respected as a man who knew how to energize his research team, to get everyone to contribute and keep them happy.

Longtime associate Bob Wentorf remembers Nerad's unique style. "He knew that a man could not do his best if he were pestered with red tape or disturbed by rumors. He shielded us from such distractions. He knew that a man could do better work if someone else took an interest in his progress. He would visit each of us at least twice a week, often more, and sit down and talk about what we were trying to do." And finally, Wentorf recalls, "He argued that if you weren't having fun, no matter what the circumstances, it was your own fault. Don't go around complaining – it's a sign of incompetence."[3]

Tony Nerad jumped at the chance to direct GE's diamond-making effort, and he began to assemble his research team. First on board were Francis Bundy and Herbert Strong, two eager and experienced scientists with much in common.[4,5] Both men were in their early forties, both received their physics Ph.D.'s from Ohio State, and both were five-year veterans of GE, where they fostered a fierce corporate loyalty that persists to this day. A close friendship developed between the Bundy and Strong families, who spent many vacations hiking and camping together and shared the excitement and occasional danger of their favorite hobby, gliding.

For the several years Bundy and Strong shared a top-floor office in General Electric's historic Building 37, the neat red brick office with the famous bold fluorescent GENERAL ELECTRIC sign and circular GE logo (their office was just below the "EL"). As members of Nerad's Mechanical Investigation Section, Bundy and Strong had been trying to improve the efficiency of the insulation used in GE refrigerators. Heat is relentlessly opportunistic, and it seeks out cold places, flowing along every available pathway to even out the temperature. It is a law of physics that no matter how hard you try, you can never remove all heat from an object, and the colder you want to make something, the more energy you must spend. That's why you have to plug in your refrigerator and pay electric bills.

Bundy and Strong knew that heat flows best along solid pathways, so they reasoned that the best insulation would have the fewest solid pathways. Their simple and elegant solution was vacuum-encased fiberglass, in which randomly criss-crossed glass strands have minuscule contacts, but also fill space. Thus they had created one of the best insulations ever seen. The only drawback was the need to encase their insulation in a vacuum-tight enclosure. Reliable vacuum-tight metal panels proved too heavy for a domestic appliance. Plastic panels might have provided a good alternative, but no one had yet developed a reliable vacuum-tight plastic. Their superinsulation has been used in a few specialized situations, such as vacuum bottles that hold liquefied air, but so far that is about it.

In spite of their parallel lives and joint research, Bundy and Strong came to their careers in science by very different routes. Francis Bundy, the son of a greenhouse farmer, was born just north of Columbus, Ohio. As an undergraduate at Otterbein College he majored in physics, and continued his education as a graduate student at Ohio State. Following graduation, Bundy joined the Physics Department at Ohio University in Athens and completed his first popular research paper, a detailed study of the physics of falling chimneys, soon thereafter. When a masonry chimney topples over, it invariably buckles and breaks somewhere near the middle during the fall. Bundy showed how to predict the details of the phenomenon, and one of his dramatic photographs of a collapsing brick tower graced the front cover of the prestigious *Journal of Applied Physics* when the article appeared in February, 1940.

He was an early bird, typically arriving at work by 7.00 a.m., before the others, and often spending lunch alone in his office with a sandwich and a recent journal. Francis Bundy says that he never went to GE to get rich. "I don't have any idea what I'd do with a million dollars," he muses. "I did the work because it was interesting and because it was useful to others." Nor did he seek the fame that many scientists crave; he never worried that most of his proprietary ideas could never be published in scientific journals.

Herbert Strong was also Ohio born, the son of a piano tuner who was deeply committed to the Jehovah's Witnesses. "I never liked it," Strong remembers of his rigid religious training, and, in turn, his parents didn't have much sympathy for Herb's fascination with science. The would-be scholar graduated from Wooster High School in 1926, but he was resigned to the fact that college would never be in the picture – that is, until his brother-in-law offered to support his studies at Toledo University. Strong majored in chemistry and physics, graduated top of his class, and never looked back. It was on to Ohio State and a physics Ph.D. in 1935.

Francis Bundy and Herb Strong met in graduate school at Ohio State in 1932 and have remained fast friends ever since. After school Strong went to work on adhesives at the Kendall Company in Chicago, but at the instigation of Bundy, he joined GE in November of 1946. A close friendship developed between the Bundy and Strong families. They frequently went camping, canoeing, or on skiing trips together. The Strongs had a cabin on Jenny Lake, an undeveloped site forty miles north of Schenectady in the Adirondacks, and it became a favorite spot for weekend getaways. These outings mixed business with pleasure, for they gave Francis and Herb much time to talk about research ideas.

And then there was the gliding. For years Francis Bundy had wanted to fly. He recalled his boyhood fascination with turkey vultures, those great soaring scavengers who glide for hours. In 1929 he formed a club with his father and some friends to build a glider, but financial constraints of the Great Depression eventually put an end to that dream. A decade later, he planned a vacation to learn to fly, but the Japanese attack on Pearl Harbor in December, 1941, changed a lot of people's plans, and the flying lessons once again had to wait. Bundy went to work on war-related research at the Harvard Underwater Sound Laboratory.

It was not until the General Electric years that Bundy's old dream came true. Francis and his son learned to fly gliders in 1952, and within a couple years Mrs. Bundy, Herb Strong, and Strong's sons were also hooked on gliding.

Soaring evokes an image of complete freedom – letting the wind take you where it will. But that is not the way for serious glider pilots like Francis Bundy and Herb Strong. For them there were skills to acquire and goals to attain. Gliding is a scientific art, requiring skilled reflexes and informed analysis of wind and weather; an act of glorious freedom in which little is left to chance.

In a way, gliding was a lot like their approach to diamond making. It was an art to be mastered – reaching for the highest temperatures and pressures, always pushing the limits of their equipment, always maintaining control, always thinking ahead.

* * *

For Bundy and Strong, their 1951 commitment to Tony Nerad and the diamond-making project involved more than just a change of research. The effort was to take place in a completely new General Electric research facility, The Knolls, which had opened in 1948. GE acquired the grand estate overlooking the Mohawk River from the Hanson family, which had made its fortune from "Pink pills for pale people." Francis Bundy, for one, was delighted with the change. His home was just a short drive from the new laboratory, but he had another commuting option. On days when the weather and his inclination were favorable, he would canoe the two miles down the Mohawk River and simply walk up the knoll to work.

The GE high-pressure laboratory was situated near a loading dock on the new building's ground floor. It was also right next to the extensive GE machine shop, a fortuitous circumstance considering how many parts would be smashed and broken in the coming years. The new facility quickly began to take on a character all its own, thanks in part to an executive assistant, Dudley Chambers, who selected a truly awful shade of institutional green to coat just about everything. "Chambers green" became a trade mark of The Knolls, and the color persists to this day.

How do you begin making diamonds? All scientific progress builds on the past, so Bundy and Strong did what any good researchers would do first. They scoured the literature, tracking down classic papers by Hannay, Moissan, and others. But mostly what they found were stacks of articles by Percy Bridgman.

Any new project at GE required at least a little paperwork, in the form of a research proposal, to define objectives and methods. In June, 1951, Bundy, Strong, and Nerad accordingly prepared a short document in which they briefly spelled out the two major problems they needed to overcome to synthesize diamond. First, the GE team had to devise a method to sustain the high temperatures and pressures at which diamond might form. Bridgman's designs seemed a good starting point. Then, they had to find a chemical in which graphite would dissolve and out of which diamond would crystallize. From the beginning they placed considerable emphasis on iron, which was known to dissolve carbon easily.

You can't do high-pressure research without a press, so Bundy and Strong scrounged up an antiquated four-hundred-ton hydraulic press that dated from the turn of the century, when General Electric was still a fledgling concern. The press was evidently used in GE's earliest experiments on extruding tungsten filaments for light bulbs. These days when people talk about hydraulics, they are usually referring to devices that use a sophisticated synthetic oil in the pressure lines. But this old wreck was a true *hydraulic* press that ran on water pressure. "It leaked so badly," one of the team members recalls, "that rubber footwear, mop, and bucket were standard accessory equipment, and the press's hydraulic lines were wrapped with rags to reduce the overhead water spray." Even so, four hundred tons was better than nothing.

Bridgman's flat anvil apparatus, in which the sample is squeezed between two tough Carboloy surfaces, held all the high-pressure records in 1950, so Bundy and Strong quickly decided to start there.

Bundy travelled to Cambridge, Massachusetts, where he was cordially received by the grand old man and given a comprehensive tour of his laboratory, advice on building and operating high-pressure devices, and sets of plans for his home-grown apparatus. Following Bridgman's blueprints, the GE team constructed carbide anvils with supporting steel rings, and they adapted Bridgman's technique of employing an "O"-shaped ring of Indian pipestone as a gasket to confine the sample. The trouble was that only Percy Bridgman seemed to have a reliable supply of the exotic material.

Bridgman's high-pressure career succeeded for many reasons: creative designs, hard work, and being in the right place at the right time. But he also gave more than a little credit to the lucky breaks. Indian pipestone – a natural gasketing material that provided both the necessary deformation and strength – was one of those breaks. In 1950, the only satisfactory high-pressure gasketing material was high-grade Indian pipestone from quarries in Minnesota, and it was only found on a Native American reservation.

In desperation, Bundy and Strong wrote to Bridgman in the hopes he might be able to help them out. Bridgman's answer was typically to the point: he told them, "If you want pipestone, you have to know an Indian." And being of generous disposition, he supplied them with a possible name (but not that of his own carefully guarded source).

The General Electric Company wrote a clear, detailed letter to this man, requesting his help in acquiring a reliable supply. In due course a rambling, four-page, hand-scrawled letter came back. The semi-literate correspondent complained about the government, the weather, and his mother-in-law. Finally, in the last two sentences, he wrote: "Have pipestone. Send $80."

GE sent the money. The pipestone never came.

For their first experiments Bundy and Strong were forced to use an inferior grade of pipestone, and their first results in November 1951 were hardly encouraging. In their naivety they tried to grow diamond simply and directly by passing high electric current pulses through compressed graphite disks. Bundy's lab notebook of November 23rd, 1951, captured a typical effort: "The test was run between 9.30 and 11.00 a.m. – The sample was loaded to 150 000 atmospheres and 20 bursts of heating current were passed through it. The calculated temperature was 1400°C.

"Then the load on the press was gradually decreased. The Carboloy anvils snapped and spalled (fragmented) as the load was diminished. When the upper anvil was lifted clear, the pipestone was scattered around with many flaky fragments of Carboloy, but the graphite sample remained complete as a nice one-centimeter diameter disk on the bottom anvil. The top surface of the anvil fell away completely."

There was not the slightest hint of transformation in the sample. It was time to get serious.

* * *

Tony Nerad asked for better equipment and more manpower. First, they urgently needed a better press. A rough calculation suggested that a 5000–ton press would be optimal, but such a machine would be huge and prohibitively expensive. After much debate, they settled on a 1000–ton design, weighing 55 tons and standing two stories tall. Blueprints of the beast were drafted and sent out for competitive bids. The Birdsboro Company in Pennsylvania made the best offer – about $125 000. A purchase of that magnitude required corporate-level approval, so a second internal proposal titled "Exploratory Project in High-Pressure and High-Temperature Processes" was submitted to the management. Diamond was certainly the prime goal on everyone's mind, but scientists can't always promise specific results. Guy Suits, Vice President of research, took the view that discovering a process to make diamond would be the ideal outcome, but something good was bound to turn up in any case. The expenditure was approved.

With the failure of their Indian contact, the diamond-making team needed to locate a reliable source of gasketing material. They took this problem to a General Electric expert in ceramics, Louis Navias, who suggested the mineral pyrophyllite, a soft, machineable rock sold commercially as "wonderstone." Wonderstone, with its extremely uniform properties and high-temperature stability, was perfect (it's still used by high-pressure workers around the world), and it saved the day. The irony, not lost on GE workers, was that the best source of wonderstone for diamond making is South Africa.

New presses and better materials were important, but Tony Nerad knew that the real key to success was the right personnel. In late 1951 he added a much-needed technician, James E. Cheney, who had recently graduated from Siena College with a degree in biology. Carrying the official title of Engineer, Jim Cheney's job was to assist the others, especially Herb Strong, in the innumerable tasks associated with the complex experimental program.

Also on board in 1951 was Harold P. Bovenkerk, who was to become a central player in the commercialization of synthetic diamond. Bovenkerk, a graduate of the University of Michigan, went to GE as an Engineer in 1947 right after serving the better part of four years in the Air Force. Bright, affable, eager, and devoted to his company, Bovenkerk knew he wanted to be an engineer from the time at age twelve, when he took apart a friend's car and successfully put it back together without him ever knowing. Most junior GE employees were bounced around from project to project, three months at a time, but Francis Bundy and Herb Strong knew a good man when they saw him, and Hal Bovenkerk was too good to let go. They convinced Tony Nerad to keep him on as a member of the team.

Bovenkerk had originally joined Bundy and Strong to work on insulation; he continued that work after Bundy and Strong began their high-pressure studies, but found himself increasingly drawn into the fascinating challenge of making diamonds. In the process, the young bachelor became almost like another son in the Bundy and Strong households. They spent many vacations together, and Hal became a glider enthusiast as well.

About the same time Tony Nerad, who recognized the critical role of chemistry in their synthesis project, gathered twenty members of the General Electric Chemistry Division together and asked if anyone wanted to join Project Superpressure. Only one young scientist, H. Tracy Hall, accepted the invitation. He went on to become a key player in the diamond-making team.[6]

Hall stands out as different from the rest of the original GE diamond makers. He was intense, brilliant, driven, and more than a little egocentric, and it's easy to see how frictions might have arisen between him and his coworkers. Unlike the others, Hall saw his own role in the diamond story as central; unlike the others he came to feel betrayed by an ungrateful company. His distress was all the more difficult to bear because, for as long as he could remember, Hall had wanted to work for General Electric.

As a boy, Hall was fascinated by the company that grew out of Thomas Edison's inventions and transformed the American landscape with electric power. Once a week his parents would go into town, dropping Tracy at the Ogden, Utah, town library. He would devour books on science and technology; Henry Ford and Thomas Edison were his heroes. When his fourth grade teacher asked students what they wanted to do when they grew up, without hesitation Hall answered, "I want to be a scientist at GE."

But the dream was not easy to fulfill. Hall was born in Ogden, Utah, son of a devout Mormon, descendant of Utah pioneers of the early 1850s. At the depths of the depression, when Tracy was only seven, his family purchased a five-acre farm a few miles outside of town. The timing of the move was terrible and the Hall family, unable to pay the mortgage, slipped deeper and deeper in debt. Though Tracy was an excellent student, his family was in desperate financial trouble and college was by no means assured. Perhaps because of those vivid child-hood memories money has remained important to Hall. His conversations were always peppered with references to prices and salaries and investments that should have paid off.

Through sheer hard work and determination Tracy Hall managed to attend local Weber College for two years, spent a year working for the United States Bureau of Mines, and resumed studies at the University of Utah in Salt Lake City. Eventually, Hall's Ph.D. in physical chemistry was awarded in August, 1948. Two months later Tracy Hall fulfilled his dream and went to work at General Electric's research laboratories.

His employment apparently began with mixed signals. He later wrote: "I found this company to be disinterested in acquiring my services. I was persistent in seeking employment, however, and was hired – reluctantly – by the G. E. Research Laboratory in the fall of 1948."

Hall joined General Electric's Chemistry Division and was assigned to work on new plastic-like materials. Du Pont's remarkable Teflon had just come on the market, and GE worked feverishly to develop rival materials with similar properties. Hall's first exposure to high-pressure research came at this time. General Electric had designed a promising new plastic, but it wouldn't dissolve in any known liquid (plastic is only useful if it can be dissolved and poured into molds). Hall knew that high pressure increased the dissolving power of many chemicals, and he found a number of solvents that worked effectively at twenty or thirty atmospheres pressure. Management had little interest in Hall's discoveries, however, and he was not entirely sanguine about his future prospects as a GE chemist.

It was in this uncertain frame of mind that Tracy Hall heard Tony Nerad's call for volunteers. Hall had dabbled in the fascinating diamond-making problem at the University of Utah, where he studied the effects of strong electric fields on carbon crystallization. He jumped at the chance to participate in the expanded G E venture.

* * *

Shortly after Tracy Hall's joined Project Superpressure, the final member of the original diamond-making team came on board (fig. 26). On the last day of 1951, Robert H. Wentorf, Jr, twenty-five years old and fresh out of graduate school, arrived at General Electric.[7]

Wentorf was no stranger to the world of corporate research and development. He was born and raised in West Bend, Wisconsin, where his father was a design engineer for the West Bend Aluminum Company, manufacturer of aluminum pots and pans and numerous other household goods. Bob Wentorf was a natural-born experimenter. Before the age of five he delighted at watching sparks fly and lights go off as he hooked up a string of metal objects and plugged both ends into the wall

Fig. 26 The General Electric diamond makers, *circa* 1955. From left to right: Francis Bundy, Herbert Strong, Tracy Hall, Robert Wentorf, Anthony Nerad, and James Cheney. (Courtesy of F. R. Boyd.)

socket. "My father had to lock up the tool chest to keep me out of trouble," he recalls with a smile.

Wentorf enrolled in the chemical engineering department at the University of Wisconsin in 1944, but he found his interests lay in the more theoretical aspects of chemistry, so he switched to physical chemistry in 1948 for his Ph.D. studies. He was always thinking about his research and always had new ideas. "They often came to me while I was shaving. Most of them didn't work," he is quick to add.

Upon graduating in late 1951 he went straight to GE, settling in Schenectady with his wife. Having inherited a considerable fortune, he didn't need to work for a living; apparently, he did it just for fun. The only obvious sign of Wentorf's wealth was his sleek black Porsche. Tracy Hall recalls one wild eighty mile-per-hour ride on the newly opened, virtually deserted New York Thruway. In those days you had to buy a twelve dollar permit to drive on it, and Wentorf was one of the privileged few who could afford one.

Wentorf was immediately perceived by his colleagues to be very bright, able, soft spoken, and the kind of guy you liked to be around – traits he retains to this day. He is a tall, wiry man, with twinkling eyes and a hint of a smile – a Norman Rockwell kind of figure. His large, strong hand envelops yours as he greets you, and with his easygoing manner he always made friends quickly. He shared an office with Tracy Hall (fig. 27) and remembers with pleasure their daily conversations about diamonds and life.

In 1952, Wentorf and Herb Strong discovered their mutual love of swimming, and they began to use the nearby, under-used Union College pool. After a vigorous swim the two would get to talking about science. "Quite a lot of progress was made in the shower room," Wentorf recalls. Bob Wentorf also gladly accepted Francis Bundy's and Herb Strong's invitation to join the glider club and he became an avid pilot himself.

* * *

Project Superpressure was to be carried out in secrecy. Only a few people were supposed to know the purpose of the giant press and the intense pace of the high-pressure researchers. Francis Bundy and Herb Strong acknowledge that their wives and children were well aware of the project, but their families also knew enough to keep quiet about it. Was

Fig. 27 Tracy Hall (seated) and Robert Wentorf, Jr., at General Electric, *circa* 1954.
(Courtesy of Robert Wentorf, Jr.)

Project Superpressure well known among other General Electric researchers? "No one was *supposed* to know," Herb Strong declares, and Bob Wentorf winks.

When scientists attempt to do something that no one has done before, they have to be willing to try lots of new ideas, most of which prove to be wrong. "You make mistakes as fast as possible," Bob Wentorf quips, "but try not to make the same mistake twice." There was no way for Tony Nerad to legislate success, but there had to be progress on two broad fronts. First, they needed a better device for obtaining and sustaining high pressure and temperature – preferably one that didn't break every time it was used. They also had to solve the complex chemical problem of diamond synthesis, because direct conversion of graphite didn't seem to work. All of the team members – Bovenkerk, Bundy, Hall, Strong, and Wentorf – were free to take off on their own, to try their hunches and take chances. At the same time, all were encouraged to interact and share ideas.

Tony Nerad decided that it was important to have a supply of diamonds to test and use as seed crystals to induce diamond growth. On May 16, 1952, Tracy Hall spent a day in New York City at the Acme Diamond Tool Company, where he was shown the bewildering variety of diamonds used in industry: gem-like crystals with perfect facets for coarse grinding; tiny shard-like diamond chips, called points, obtained from rough shaping of gems; and black masses of microscopic diamond. Hall was authorized to purchase up to $1000 worth of industrial diamond; his actual bill totaled $994.72, almost half of which was for 19 carats (about a tenth of an ounce) of diamond points – specimens that would later play a central role in the intense controversy over who made the first diamonds.[8]

Francis Bundy came up with the group's first new high-pressure device, a modification of Bridgman's anvils dubbed the "flying saucer" because of the distinctive shape of pressurized samples. Bridgman-type anvils failed because high temperatures softened and eroded the Carboloy surfaces. Bundy circumvented the problem by altering the anvil shape and using protective inserts of magnesium oxide for insulation. The flying saucer (fig. 28) could repeatedly stand higher temperatures, up to 2700°C over and over again, and it became the workhorse during many experiments in 1952.

Unfortunately, the saucer device was severely limited because it could not reach pressures high enough to make diamonds. Achieving high pressure requires squeezing a sample into a smaller and smaller volume, but no matter how the GE workers arranged their gaskets and samples, the two carbide anvils came into contact with each other at relatively low pressures. Once contact was made, no further sample compression, and thus no increase in pressure, was possible. At the time, Bundy believed that the flying saucer's highest possible pressure was only about 35000 atmospheres – adequate for learning the high-pres-

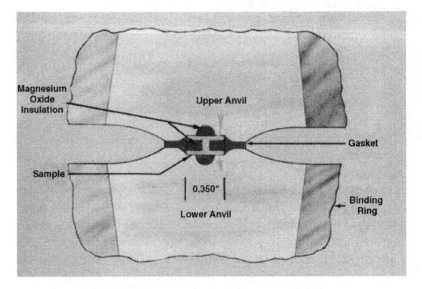

Fig. 28 Francis Bundy's "flying saucer" device, a modification of Percy Bridgman's opposed anvil apparatus, got its name from the distinctive shape of the sample chamber. (Courtesy of F. R. Boyd.)

sure ropes, but not enough to make diamonds. The team was able to convert diamond partially to graphite at high pressure – a process that hinted at the conditions needed to trigger the reverse process – but they were still a long way from their goal.

There were new devices to design and complex chemical experiments to try. It was a free-for-all, with the scientists involved in countless brainstorming sessions, each member of the team following up on his own hunches.

While this approach would ultimately produce success, it also sowed the seeds of an enduring and bitter controversy.

Notes

1. C. Guy Suits, "The synthesis of diamonds – A case history in modern science." Schenectady, NY: General Electric, 24 pp., 1960. This important corporate version of the diamond synthesis story was reprinted in Suits's collected essays, *Speaking of Research* (NY: Wiley), pp. 312–332. The quote appears on p. 312.
2. Many of the details from this and subsequent chapters on diamond synthesis at General Electric were obtained through extensive interviews and correspondence

with all of the original GE researchers: Hal Bovenkerk, Francis Bundy, Tracy Hall, Herb Strong, and Robert Wentorf. Additional information was obtained through interviews and correspondence with Francis Boyd, William Cordier, Robert DeVries, Armando Giardini, Clifford L. Spiro, Alvin Van Valkenburg, and Hatten S. Yoder, Jr.

In addition to the Suits's reference above, several historical accounts review the GE effort to synthesize diamonds. These include: Gordon Davies, *Diamond.* (Bristol: Adam Hilger, 1984); K. Nassau and J. Nassau, "The history and present status of synthetic diamond." *Lapidary Journal* **32**, 76–96, 490–508 (1978); Judith Milledge, "Natural and synthetic diamonds." *The Times Science Review*, Spring 1960, 8–13.
3. Robert H. Wentorf Jr., "The synthesis of diamonds." Unpublished typescript, 15 pp. (1989); "RE Tony Nerad" (A eulogy) Unpublished typescript, 2 pp. (January 6, 1980).
4. In addition to the references cited above, information on Francis Bundy's career is provided by his unpublished manuscript, "Diamond synthesis," prepared as a talk for the Ohio Section of the American Physical Society, 19 April 1991.
5. Some of Herbert Strong's recollections appear in: "Early diamond making at General Electric." *American Journal of Physics* **57**, 794–802 (1989), and in "Early diamond making revisited," General Electric Technical Information Series Class 1, 48 pp.
6. H. Tracy Hall, "Personal experiences in high pressure." *The Chemist* July 1970, 276–279; see also H. Tracy Hall, "The transformation of graphite into diamond." *Newsletter of the American Association for Crystal Growth* **16**, 2–4 (1986).
7. Robert H. Wentorf Jr., "The synthesis of diamonds," op. cit (note 3).
8. H. Tracy Hall, "A short report of a diamond-buying trip," GE Mechanical Investigations Section Memo Report C-90, May, 1952.

CHAPTER 7

BREAKTHROUGH

The evidence at hand confirms that the birthday of diamond synthesis was during the night of 8–9 December 1954. Very shortly afterwards, diamond synthesis became a routine procedure.[1]

HERB STRONG, "EARLY DIAMOND MAKING AT GENERAL ELECTRIC," 1989.

On December 16, 1954, I discovered how to make diamonds. Others have claimed prior discovery . . . but there was something unique about mine. My method could be reproduced by others.[2]

TRACY HALL, "PERSONAL EXPERIENCES IN HIGH PRESSURE," 1970.

IN THE FOUR DECADES SINCE GE scientists first made diamonds, two contradictory interpretations of the historic events of 1953 and 1954 have persisted. These rival descriptions, each passionately argued by the central parties and each endorsed by a loyal following, differ hardly at all in the framework of facts. But the central question – who was the first to make diamonds – long remained a point of contention.

One version of the history – the version endorsed by General Electric's management as well as by Francis Bundy, Herb Strong, and Bob Wentorf – emphasizes the years of teamwork and corporate support. Guy Suits, Vice President and Director of Research at GE, was responsible for the most extensive telling of this tale. He based his corporate history on a detailed review of laboratory notebooks, and published his interpretation in 1960 as the pamphlet *The Synthesis of Diamond – A Case History in Modern Science*.[3]

Suits provides a straightforward, upbeat narrative that underscores the crucial importance of collaborative research along GE's inexorable path to success. "I find it difficult to imagine that a single investigator could have carried the work through to a successful conclusion," he wrote. There were too many false leads, too many technical subspecialties required, and too much basic data for any one person to collect. Furthermore, he argued, "Joint effort is an almost indispensable ingredient of morale and courage." Even if one person makes a key finding, everyone in the team must share the credit equally, he argued, because "negative results contribute heavily to the final positive success." Suits recognized that this corporate policy carried with it "some penalties." In particular, "credit for progress is shared, and individual identification is partly subordinate to group achievement. But if aggregate achievement is great enough, as in this case, there is credit enough for everyone." Suits's research philosophy affected his policy for rewarding employee's discoveries, as well. Individual creativity was viewed as secondary to group effort and all members of a team shared in the credit and financial reward.

Suits's corporate history is echoed and amplified by Herb Strong's two more recent articles, "Early diamond making at General Electric"[4] and "Early diamond making revisited,"[5] which examine in particular the critical events of November and December, 1954. Strong, too, speaks proudly of GE diamond making as a paradigm of cooperative research and unselfish collaboration. Over the years the story has become a kind of company gospel, retold in countless brochures and even cast in bronze at the entryway of the main diamond-making plant in Worthington, Ohio.

However, a very different version of the history comes from the pen of Tracy Hall, who argues for his own unique and pivotal role in the synthesis of diamond. Hall publicized his interpretation in several short, bitter articles, including the outspoken "Personal Experiences in High Pressure" that appeared in *The Chemist* in July of 1970.[6] Hall dismisses the importance of collaborative research at GE, and promotes his own central role. Though hampered by a bureaucratic management and unsympathetic coworkers, his story goes, Hall solved the key problems of device design and chemistry almost entirely by himself. Many in the scientific community, impressed with Hall's brilliant subsequent innovations in high pressure, have tended to accept this version as closer to the truth. Tracy Hall is unequivocally cited as the first to make diamonds in several authoritative histories, and it was he alone who received the American Chemical Society's gold medal for creative invention in 1972.

As Hall's narrative implies, the history of General Electric diamond making is actually several stories, complexly intertwined. Bovenkerk, Bundy, Hall, Strong, and Wentorf were all gifted men; each had his own ideas and ambitions, and each followed his own hunches. The General

Electric history is the story of a team, constantly brainstorming and constantly catalyzed by Tony Nerad. The extent to which ideas were developed collectively versus independently is unrecorded, but there was certainly a great deal of interaction. Nerad held group meetings at least twice a week to hash out ideas and discuss priorities, and there were incessant informal conversations as well. Still, by everyone's account, Tracy Hall was the odd man out.

In Hall's view, the reason for his isolation boiled down to one simple fact – he was a Mormon, and that didn't sit well with the others. His social life revolved around the church and his large family. Weekend camping trips, gliding, skiing vacations, and socializing after work with the other guys were out of the question. "My problem was that the others were a club," he laments. But the isolation seemed to go beyond that. Hall is convinced that he was the victim of religious discrimination. General Electric promotions and raises were based in part on peer review; Hall believes that he was penalized for spending too much time with his church. And he remembers the jokes – vicious, distasteful jokes about Mormons and polygamy. How much of this is true (the others strongly deny any prejudice), and how much was in the imagination of an embittered employee, is impossible to say, but the rift was real.

Money represents another bone of contention. For Francis Bundy, Herb Strong, and Bob Wentorf promotions, raises, and bonuses never seemed a central concern. Bundy and Strong lived simply, had small families, and were content to spend vacations camping in a rustic setting. Wentorf inherited a small fortune from his parents and was unconcerned about salaries and retirement. But Hall had a large family to support and a church that expected sizeable contributions as well. Tracy Hall desperately wanted to improve his family's lot, and he expected much more from General Electric than he ever got. "I went five years without a raise," he laments. When he made diamonds he was finally given a paltry salary increase, from $10 000 to $11 000 – "much less than what Bundy and Strong were making. I loved GE, and they didn't love me back." Hall speaks passionately about his role: "I hit the home runs, but they took the credit. It was worth a Nobel." Tracy Hall has found many reasons to be resentful.

Who was the first to synthesize diamond? The answer lies not so much in what any one person did on a particular day, but rather in the nature of scientific discovery, itself. What does it mean to make a discovery? What does it mean to be first?

* * *

Early in 1953 the GE diamond makers faced a major dilemma.7 Bundy's flying saucer device could sustain high temperatures, and so was a distinct improvement over Bridgman's simple opposed anvils. But the saucer couldn't compress samples sufficiently to generate really high pressures. It was brainstorming time; everyone was pushed to think about new devices.

Piston-in-cylinder designs provided one promising alternative. A simple piston-in-cylinder arrangement can, at least in priciple, achieve any desired pressure by squeezing a sample into a smaller and smaller volume. The piston-cylinder is limited, however, by the difficulty of supporting the piston, which tends to break under the stress of high temperature and pressure. The GE team had to find a compromise, combining the durability and temperature capabilities of the saucer with the ease and pressure capabilities of a piston-cylinder device.

Strong, Wentorf, and Hall all designed and tested modifications of the basic piston-cylinder device. Each scientist drafted plans, waited impatiently for the machine shop to build the device, and then waited some more for time on the press to try it out. The men came up with all sorts of intriguing possibilities like the "stubby piston and cylinder," the "collared piston and cylinder," and more. Wentorf was especially prolific, though none of his devices was very successful. "I have more high-pressure apparatus patents than Herb and Francis put together, and none of them ever worked," Wentorf laughs.

Attaining high pressure was only part of the battle. They also had to heat their samples to temperatures approaching 1500°C. Strong and Bundy favored the tried-and-true technique of passing electricity through a coil of platinum wire, which glowed intensely hot around the sample. The technique worked, but replacing the platinum coil after each run was expensive and time consuming.

The GE team also tried heating samples by passing electric current through a graphite heater, much as Loring Coes had done. This simple and elegant solution to the high-temperature problem used a short graphite cylinder to surround the sample. The graphite heater, furthermore, provided more carbon for potential conversion to diamond. Each of the different device designs took a slightly different graphite sample assembly, but it was an easy matter to shape the heater as required.

As competition intensified and the queue for machining and press time lengthened, the first real signs of conflict began to disrupt the group. During 1953 Herb Strong and Tracy Hall each developed a device that combined many of the best features of the saucer and piston-cylinder designs. Strong's "cone apparatus," the more conservative of the two machines, featured two simple beveled anvils with flat surfaces, much like Bridgman's, but also incorporated a thick-walled cylinder of steel or carbide to provide a larger sample chamber. The upper anvil drove a Carboloy piston into the sample chamber (fig. 29).

While teamwork was the GE party line, individual initiative often proved the best path to success. At times this reality has led to conflicting versions of the history; the gasketing procedure for the cone apparatus is a case in point. Rather than use a single wonderstone gasket, Strong's device relied on a clever sandwich gasket of alternating layers of wonderstone and soft steel. This arrangement could accomodate larger samples and sustain higher pressures. Tracy Hall claims that *he* suggested this improvement, which set pressure records at the GE lab. Hall recalls that his idea of a sandwich gasket was rejected when first proposed at a group meeting but was adopted by Strong a short time later, without credit. Hal Bovenkerk, on the other hand, believes that he was the first to suggest the modification, while Herb Strong insists that the sandwich idea was an obvious way to increase pressure; such composite seals had been used in high-pressure work for more than half a century. Whatever the truth, the cone apparatus was far more successful than previous GE devices. It easily reached 40 000 atmospheres at temperatures up to 2000°C.

Hall's rival device, the "semi-piston and cylinder" or "half-belt," displayed many of the same features as Strong's cone apparatus but it represented a much more radical departure from traditional high-pressure designs. The upper anvil had an unprecedented graceful concave surface that fit into a well in the lower anvil, which had a slightly steeper curve (fig. 30). This complex geometry reduced piston failure, helped to control gasket failure, and provided a greater sample compression than the saucer, while maintaining the saucer's heating capabilities. Hall was able to push his half-belt repeatedly to 40 000 atmospheres and 2000°C – the temperature at which the device itself melted.

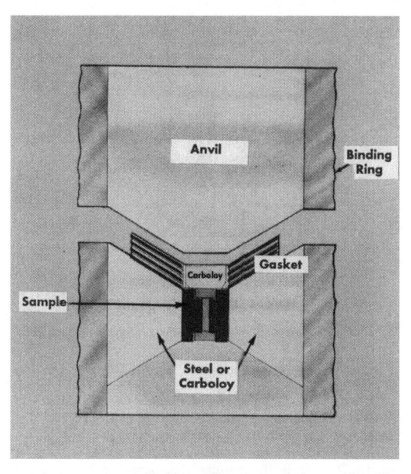

Fig. 29 Herb Strong's cone apparatus featured a strong binding ring and a layered gasket of alternating steel and pyrophyllite cones that helped to confine the sample laterally as it was compressed by the massive piston. (Courtesy of F. R. Boyd.)

With three working devices – Bundy's flying saucer, Strong's cone apparatus, and Hall's half-belt – and only one press (the 1000–ton press was not expected until early 1954), there were bound to be conflicts. Hall was convinced that his half-belt idea was best, and Strong was equally determined to promote his cone apparatus. These tensions were exacerbated when Tracy Hall asked to build an even more radical device – the full belt. Hall complained: "The Half-Belt gave higher steady-state pressures and temperatures than had ever before been achieved simultane-

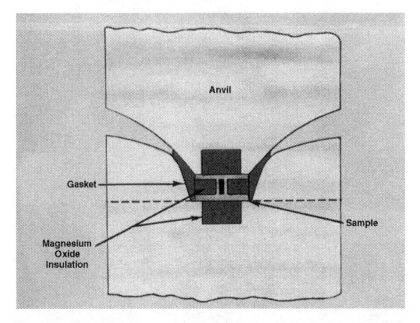

Fig. 30 Tracy Hall's half-belt apparatus incorporated features of both the flying saucer and the cone. However, the gradually tapered piston and curved pyrophyllite gasket was original to this design. (Courtesy of F. R. Boyd.)

ously. But because my colleagues felt negatively about it, when I proposed to build an improved version, the 'Full-Belt' or just plain 'Belt,' the proposal was rejected, although the cost was less than a thousand dollars. I fretted about this for a time and then decided on a sub-rosa solution. Friends in the machine shop agreed to build the Belt, unofficially, on slack time. This took several months. Ordinarily, it would have taken only a week."[8]

The completed belt apparatus was a thing of beauty – sleek and shiny with curves of elegant function and form. Hall had relied on the traditional opposed piston geometry, but he tapered and curved his pistons and drove them into the rounded openings of a steel doughnut – the "belt" (fig. 31). As the two tapered anvils compress the ends of the disk-shaped sample with increasing force, the belt provides more support around the circumference of the sample. Hall also relied on clever wonderstone and soft steel gaskets – called "flowerpots" because of their distinctive tapered form – to hold the sample in place.

Hall's simple, elegant belt apparatus usually worked wonderfully

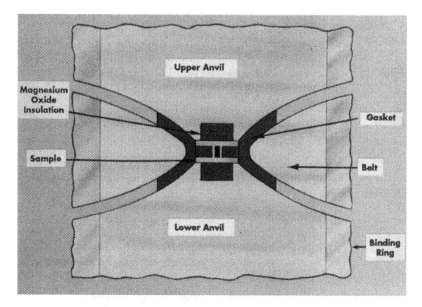

Fig. 31 Hall's full-belt apparatus employed two tapered pistons and a doughnut-shaped binding ring. This design greatly increased the stroke – the total travel distance of the pistons – and thus allowed for higher pressures. (Courtesy of F. R. Boyd.)

well, though the tapered pistons had an annoying habit of breaking after every few runs. The belt was easy to use with a graphite heater, and extreme temperatures and pressures could be maintained for long periods without difficulty. "It operated so successfully, in my view, that I desired to have the critical components constructed of [Carboloy]. This would allow much higher pressures to be generated. Management, however, would not approve the purchase of the carbide."

In spite of General Electric's expertise in making carbide components, the superpressure group did not have unlimited access to the material. Carbide pieces were expensive and time-consuming to produce. Bundy's flying saucer and Strong's cone apparatus had first been built from hardened steel, and it seems to have been standard policy to test the steel version fully before going to carbide. Nevertheless, Hall saw the denial as another slight, another cause for resentment to build, and once again he looked elsewhere for support. "Having been stopped by the Superpressure people, I appealed to my former supervisor [chemist Herman Liebhafsky] and spoke at a seminar of his group concerning the Belt. He and his group were impressed and shortly thereafter permission was received to buy the carbide components."

By early 1954 the General Electric Superpressure team had its devices in place. Herb Strong secured his cone apparatus in the new two-story, thousand-ton Birdsboro press, which had just arrived at the lab. Tracy Hall's improved belt apparatus was assigned – relegated, Hall would say – to the antique four-hundred-ton hydraulic press (fig. 32), where Hall and Wentorf were able to sustain experiments at 70 000 atmospheres, with shorter excursions to pressures estimated as high as 100 000 atmospheres. The race to make diamond could proceed in earnest.

* * *

Achieving high pressures and temperatures was only one arena for the fierce competition between scientists. The men were also in a race to discovery the chemistry of diamond synthesis. Hundreds of promising avenues had to be tried, and each of the scientists was encouraged to pursue his own ideas. Chemist Bob Wentorf decided to run a series of experiments on sodium carbide and lithium carbide, and he was the first to "smell" diamond. In the fall of 1953 he found that lithium carbide decomposes at high pressure and temperature, producing a few very tiny crystals that might have been diamond. Herb Strong recorded the success in his notebook: "From many tests on scratching glass, and [x-ray evidence], we now feel quite certain that diamonds were made," though he later backed off this claim.

In spite of these initial indications, repeated attempts failed to improve the yield or clarify the x-ray results, and that tack was finally abandoned late in 1953. It is a pity that all the samples, which sat in a drawer for years, were thrown away a decade ago without ever being reanalyzed. Modern analytical techniques could easily have revealed any traces of synthetic diamonds.

Wentorf also experimented with mixtures of aluminum oxide and carbon. When heated and squeezed, the mixture produced hard, brilliant crystals, less than a thousandth of an inch across, with diamond's familiar octahedral shape. But Wentorf's hopes were dashed when further analysis showed them to be corundum, the common abrasive form of aluminum oxide. Wentorf also studied the high-pressure

Fig. 32 Francis Bundy, Herb Strong, and Jim Cheney operate the thousand-ton, two-story-tall Birdsboro press, while Tracy Hall (left center) uses the antiquated four-hundred-ton hydraulic press, *circa* 1954. (Courtesy of H. T. Hall.)

behavior of garnet, and was able to reproduce some of the changes that occur in rocks when subjected to high pressure.[9]

Tracy Hall had his own agenda. He concentrated on decomposing various carbonates (compounds containing carbon and oxygen), including the lithium–lithium carbonate system. In this strategy Hall received much encouragement from Samuel Kistler, the Dean of Engineering at the University of Utah, who had initiated Coes' diamond-making efforts at Norton a few years earlier. Hall, like Wentorf, produced tiny, hard, diamond-like crystals, but they were too small for positive identification. With the new high-pressure capabilities of the belt apparatus, Hall also tackled the challenge of direct conversion of graphite to diamond, but even at estimated sustained conditions approaching 100 000 atmospheres and 4000°C no diamonds were detected.

The belt apparatus was a workhorse and Hall and other team members used it for hundreds of experiments in 1954. Having failed in the direct conversion of graphite, Hall switched to studies of other carbon-bearing systems. These experiments, some of the few for which General Electric laboratory notebooks are available, reveal much about Hall's strategy and frame of mind. Studies of the high-pressure forms of silica mixed with carbon were successful in duplicating Coes' synthesis of coesite, but produced no diamond. "Nov. 29, 1954 ... Exam under the microscope showed that graphite crystals are present inside most of the grains! . . . This experiment has shown that [carbon] is effectively soluble in [silicon dioxide] alone! But, it precipitates as graphite."[10] Similar experiments on iron carbonate at very high pressure occupied most of the next two weeks. They also produced graphite but no diamond.

That December, Hall spent much effort on the routine chore of calibrating his thermocouples, essential for precise temperature measurement, but he also had time to muse about very unorthodox experiments. "Dec. 3, 1954 – I should measure the effect of ultra-hi-P and hi-T on bacteria. . . . Perhaps they will stand much higher temperature at a high pressure." And he displayed a gentle sense of humor in his entry for December 14th: "Just for a gag one of these days, we could 'cook' a piece of carrot or other vegetable in the 'Belt' at 100 000 atmospheres and perhaps 1000°C (or lower if need be). It should be done in a second or so. What a pressure cooker!"

Meanwhile, in November and early December, 1954, Herb Strong was putting the cone apparatus through its paces. In his first experiments he tried to induce growth on a diamond seed crystal prepared by etching. Before the experiments the seed crystal displayed sharp angular pits, but afterwards, "All traces of the pits are gone. The edges were rounded and the faces had the appearance of a contour map as though rapid growth had occurred." In retrospect, however, it seems unlikely that the cone apparatus had achieved conditions for diamond growth.

Strong also began thinking about a more pivotal set of experiments, inspired by the discovery of diamonds in some iron meteorites. In the tradition of Moissan and Parsons, he attempted to dissolve carbon into iron and other metals. His laboratory notebook of November 22, 1954, records the crux of this key idea: "The [iron-nickel] alloys should be investigated at high pressure for [melting point] and as a solvent for carbon in the liquid state. . . . I think the work will have an important bearing on dia[mond] growth in Fe or Fe and Ni alloys."

Although the would-be diamond makers didn't realize it at the time, late November of 1954 marked a crisis point in the General Electric effort. All of the original research funds allocated to Project Superpressure had been exhausted a year before. Twice Tony Nerad had gone to GE's Vice President for Research, Guy Suits, and asked for additional funds; twice the money had been provided and used up. For a third time Nerad went to Suits, hat in hand. This time Suits brought the matter up before the November meeting of GE research managers. Nerad sensed trouble.

Project Superpressure was approaching its fourth anniversary – four years of GE investment in what, to many, seemed rather abstract at best. Diamond making was a nice dream, but no one had forgotten the previous failures of Percy Bridgman and the Norton Company. When asked to discuss the benefits of continuing the high-pressure program, Nerad had nothing concrete to offer. When asked to vote on the matter, the other managers almost unanimously turned thumbs down on continued support. Fortunately for the diamond makers, General Electric was not a democracy; Guy Suits decided, for a short while longer at least, to let the high-pressure group proceed.

Time was running out. Nerad knew that his team couldn't meet a diamond-making deadline, but he hinted to Herb Strong that some kind of breakthrough was needed. Strong took the warning to heart and decided to go all out with the cone apparatus and an iron solvent.

The starting material of choice was a fine black powder known as "Steco," a commercial carbonizing powder for making steel, case hardening gears, and other metallurgical processes where carbon is added to iron. Strong and his colleagues reasoned that if the carbon in Steco is easily dispersed in metal, it would be an ideal starting material for making diamond. Two small seed diamonds were wrapped in iron foil and embedded in the steco to act as growth centers.

Strong began the critical run, Experiment 151, on the evening of Wednesday, December 8, 1954. The cone apparatus, with its angled steel-anvil, was set at an estimated 50 000 atmospheres pressure and 1250°C. While most experiments were of relatively short duration, there was always a nagging suspicion that time might be a critical factor – after all, nature seemed to take millions of years to grow diamonds. So Strong planned this overnight attempt for an unusually long 16 hours. The run was monitored by a night watchman, who every so often recorded the heater power settings displayed on a large meter. Except for some annoying fluctuations in current that raised temperatures above the preset conditions during the night, the run seemed completely ordinary.

Strong removed the specimen on the morning of December 9th and examined the products. The two seed crystals tumbled free; they had not changed at all. No sign of diamond growth was found, and he chalked it up as just another failed experiment. Strong did notice that a portion of the iron had melted into a blob at one end of his sample chamber. One of the principal motivations for the run was to determine how much carbon could be dissolved in the molten iron. He wanted to know if any reaction had taken place among the starting materials, so he sent the sample to the metallurgy division to be polished (and prepared for light microscopy) whenever they had a few free moments.

Herb Strong was totally unprepared for the message from metallurgy a week later on December 15th.

"I'm terribly sorry but I can't polish your sample. It's gouging my polishing wheel," the technician Bob Smith complained. Strong rushed to the metallurgy labs to have a look. A distinctive octahedral point protruded from the hard, metal mass of his sample.

Strong wrote his own account of the subsequent dramatic events. "The entire group gathered around to inspect that hard point. Initially there was a moment of stunned silence. Could it possibly be diamond? Finally, Tracy Hall spoke the verdict: "It must be a diamond!'"[11]

Eventually, two shard-like diamonds, one-sixteenth of an inch in the longest dimension, were separated from the sample. X-ray analysis proved beyond doubt the identity of Strong's historic crystals.

* * *

In the subsequent weeks, months, and years the members of the General Electric diamond-making team – Tracy Hall and Herb Strong, in particular – devoted an inordinate amount of time and energy debating the origin of those two tiny crystals. What is beyond debate, however, is that the very next day, on December 16, 1954, Tracy Hall performed a similar experiment in the belt. He added two diamond seed crystals to iron sulfide (a mineral associated with diamonds in meteorites) and placed this material in a cylindrical graphite heater. He also followed belt protocol by placing thin disks of tantalum metal between the sample and the belt anvils, to bring current for resistance heating. The experiment, conducted at an estimated 100 000 atmospheres and 1600°C, took only thirty-eight minutes.

Hall described what he found in vivid detail. "I broke open a sample cell after removing it from the Belt. It cleaved near the tantalum disk. Instantly, my hands began to tremble. My heart beat wildly. My knees weakened and no longer gave support. Indescribable emotion overcame me and I had to find a place to sit down!

"My eyes had caught the flashing light from dozens of triangular faces of octahedral crystals that were stuck to the tantalum and I knew that diamonds had finally been made by man."[12]

Weeks of feverish synthesis activity followed the exciting discoveries of Strong and Hall. Reproduceability is critical to the conformation of any scientific advance, and Bundy and Strong tried numerous times to duplicate Strong's feat. Tony Nerad also assigned Hal Bovenkerk the task of repeating Strong's run independently with the steel cone apparatus in the thousand-ton press.

They couldn't do it. "We wasted weeks," Bovenkerk complains. Pressures were never quite high enough, and iron carbide always appeared instead of diamond, thus effectively blocking the path to diamond growth. It was not until more than a month later, with the installation of carbide anvil components, that the cone apparatus routinely attained pressures required for diamond synthesis.

Tracy Hall and Bob Wentorf, on the other hand, repeated their synthesis with ease. Armed with the belt apparatus mounted in the old four-hundred-ton press they made diamonds over and over again – twenty times in the next two weeks. Under Hall's supervision, GE physicist Hugh Woodbury became the first person outside the Superpressure group to confirm the synthesis claim, making a successful run on December 31, 1954.

The race to make diamonds was over.

*   *   *

For years historians and the principal players have debated who was the first to synthesize diamonds. From the very beginning there were serious misgivings about the origin of Strong's two tiny crystals, supposedly created on the night of December 8th. Tracy Hall admits to having said, "It must be a diamond," when he saw a crystal embedded in Strong's experiment. But he is quick to add, "I never said it was a *synthetic* diamond. No one believed that."

Strong's inability to duplicate the diamond-making feat without substituting carbide parts for the cone apparatus' steel anvils cast great doubt on the validity of the December 8th experiment. In two detailed, complexly argued articles, Strong suggests that temperature fluctuations during the night of December 8th were just barely sufficient to create the extreme conditions for diamond growth.[13] Hal Bovenkerk, who struggled for weeks to duplicate Strong's experiment and subsequently spent four decades making diamonds for GE, is not convinced. That frustrating, wasted month, while others enjoyed the thrill of making diamonds on Hall's belt, "made me a disbeliever," he says.

Of the two small diamonds from Strong's run, only the larger survives (fig. 33). It has become an icon of GE history, lovingly protected on a special wooden plaque with a built-in magnifier and engraved plate proclaiming it to be the "First Diamond Made in GE Research Laboratory." The plaque was presented in 1955 to General Electric's CEO, Ralph J. Cordiner, to display in his office, and there was talk of transferring the display to the Smithsonian Institution. The souvenir lost some of its appeal, however, when Tracy Hall wrote a sharp letter to Cordiner and the Smithsonian Institution discrediting the crystal and claiming priority for his own synthesis. Cordiner, not wishing to become involved in the controversy, returned the display to Herb

Fig. 33 The larger of the two diamond crystals recovered from Herbert Strong's experimental run of December 8, 1954, has characteristics typical of natural diamond. (Courtesy of Herbert Strong.)

Strong, who for a time kept it in his office. Eventually, the crystal was placed in General Electric's historical display in Schenectady.

Ironically, that sole surviving crystal, enshrined as a piece of history, now provides compelling proof that no diamonds were made on the night of December 8th. During research for this book, General Electric experts agreed to reexamine the "first" synthetic diamond with diagnostic infrared absorption spectroscopy. Their results appeared in the September 2, 1993, issue of *Nature*, in a short letter, "Errors in diamond synthesis."[14] According to the GE team, Herb Strong's crystal could not possibly be a synthetic diamond. GE insiders cite a litany of anomalies. The crystal is unusually large – perhaps ten times the size of any other crystal produced during those first weeks. The elongated shape is odd, too, for most of the early crystals were more or less round in shape. Strong's diamond is water clear, while typical synthetic diamonds are yellowish; it displays atypical surface etching; its x-ray pattern lacks distinctive "satellite" spots characteristic of most man-made diamond. And, most damning, it's infrared spectrum shows distinctive features found only in natural diamond. In fact, Strong's diamond looks for all the world like the natural diamond points that Tracy Hall brought back by the hundreds from New York City. Of the original five diamond makers, only Tracy Hall refused to be a coauthor of the *Nature* letter, in protest over the forty–year delay in official recognition of his achievement.

The "First Diamond" is nothing more than a shard from a natural stone. But, in a sense, the nature and origin of that one crystal is unimportant. Whether or not Strong's experiment was easily duplicated, whether or not the diamond crystal was grown by man or was natural and inadvertently slipped into the experiment, there can be no doubt that Strong's experiment pointed Tracy Hall in the right direction. The very day after Strong's "discovery," Hall did in fact make diamonds.

Without question Tracy Hall conducted the first experimental synthesis that was easily and independently reproduced. Hall was the inventor of the belt apparatus, and he was able to make diamond over and over again with it. But Hall's experiment of December 16th was the culmination of almost four years of focused research and was influenced directly by Herb Strong's announcement of the previous day – an experiment also based on a sample with mixed iron and carbon.

Tracy Hall never worked in a vacuum, and it is unlikely he could have made diamond were it not for his Superpressure colleagues and the company that brought them together. By the same token, Tracy Hall, perhaps by virtue of being the lone wolf of the team, played a unique and pivotal role. His belt apparatus – virtually unchanged in basic features to this day – was a brilliant advance and a key to GE's success. By ignoring the critical nature of Tracy Hall's contributions, General Electric and its spokesmen have polarized and distorted the historical record.

It is sad, in a way. For more than a century diamond synthesis stood as one of the most glamorous, intractable problems in chemistry. Some of science's leading lights, Nobelists Moissan and Bridgman among them, devoted years to the quest. General Electric's effort to synthesize diamond required major advances in attaining extreme temperature and pressure, as well as the discovery of the appropriate chemical environment for diamond growth. The team's solution to the problem was the kind of breakthrough that deserved a Nobel Prize, and could have easily commanded it with the right kind of publicity. But General Electric's original team featured a half-dozen diamond makers, and Nobels can be shared by only three people. If GE had highlighted Tracy Hall's special role, and promoted his nomination for the Prize, General Electric could have heaped glory on top of their extraordinary commercial success. With such a magnificent discovery there was plenty of credit to go around.

* * *

So much history has been lost. The exhaustive laboratory notebooks – all except the critical pages from late 1954 – are unavailable for study and may have been discarded. All of the original high-pressure samples from 1951 to 1954 were stored in a drawer and eventually thrown away. As a result, some key questions will never be answered. Did the General Electric team make diamond before December, 1954? Were there tiny diamonds embedded in those tantalizing runs of 1953? Possibly, but we will never know for sure.

Nevertheless, it is clear that General Electric ultimately won the race to synthesize diamond because, in the words of Percy Bridgman, they "found out how to apply more pressure at a higher temperature and for a longer time than ever achieved before." They learned a quick, easy, reproducible way to do something no one had done before. And for the GE diamond makers, the adventure had just begun.

Notes

1. Herbert Strong, "Early diamond making at General Electric." *American Journal of Physics* 57, 794–802 (1989), p.801.

2. H. Tracy Hall, "Personal experiences in high pressure." *The Chemist* July 1970, 276–279 (1970), p.276.

3. C. Guy Suits, "The synthesis of diamonds – A case history in modern science." Schenectady, NY: General Electric, 24 p., 1960. This essay also appears in Suits's book, *Speaking of Research* (New York: Wiley), pp. 312–332.

4. Herbert Strong, "Early diamond making at General Electric," General Electric Technical Information Series Class 1, 48 pp.

5. Herbert Strong, "Early diamond making revisited," op. cit.

6. H. Tracy Hall, "Personal experiences in high pressure." *The Chemist* July 1970, 276–279; H. Tracy Hall, "The transformation of graphite into diamond." *Newsletter of the American Association for Crystal Growth* 16, 2–4 (1986). See also: Kurt Nassau, "Who was first?" *Lapidary Journal* November 1991, 87–93.

7. Many of the details in this chapter were obtained through extensive interviews and correspondence with the original GE researchers: Harold Bovenkerk, Francis Bundy, H. Tracy Hall, Herb Strong, and Robert Wentorf, Jr.

8. This and subsequent quotations on the development of the belt apparatus are from H. Tracy Hall, "Personal experiences in high pressure," op. cit., pp. 276–277. See also: H. Tracy Hall, "Ultra-high-pressure, high-temperature apparatus: the 'belt.'" *Review of Scientific Instruments* 31, 125–131 (1960).

9. Robert H. Wentorf, Jr., "The formation of Gore Mountain garnet and hornblende at high temperature and pressure." *American Journal of Science* 254, 413–419 (1956).

10. Quotations from Project Superpressure laboratory notebooks are taken from the facsimile reprint that appears in Herbert Strong, "Early diamond making at General Electric," op. cit., pp. 12–48.

11. Herbert Strong, "Early diamond making at General Electric," op. cit., p. 798.

12. H. Tracy Hall, "Personal experiences in high pressure," op. cit., p. 277.

13. Herbert Strong, "Early diamond making revisited," op. cit.; "Early diamond making at General Electric," op. cit.

14. H.P. Bovenkerk, F.P. Bundy, R.M. Chrenko, P.J. Codella, H.M. Strong, and R.H. Wentorf, Jr., "Errors in diamond synthesis." *Nature* 365, 19 (1993).

CHAPTER 8

SECRETS

In February of 1955 I must confess that an ill-timed sneeze in the wrong place would have wiped out the entire world supply of Man-Made diamonds.[1]

C. GUY SUITS, "SCIENTIFIC COMPETITION," 1963

I F ANYTHING FRIGHTENED THE General Electric diamond makers more than failure, it was the prospect of proclaiming success and then being proven wrong. The history of diamond research is littered with mistaken claims of synthesis, if not outright fraud, and General Electric managers wanted no part of the embarrassment of a bogus announce-ment. Guy Suits insisted that Tony Nerad's group redouble its efforts and take every possible step to confirm the discovery.

Confirmation depended on a three-fold strategy. First, make as much diamond as possible with Tracy Hall's proven technique. Second, subject those samples to every test in the book to confirm their identity and test their grinding properties. Third, discover as many other ways as possible to make diamonds at the high pressures and temperatures available with the belt and cone apparatus.

Tracy Hall was on a roll, synthesizing diamond in run after run in the carbide-tipped belt. Each of his fifteen- to twenty-minute experiments produced up to a quarter of a carat of tiny diamond crystals. Hall's first diamond-making experiments had used an iron sulfide, but GE workers soon discovered that iron metal was the crucial solvent and worked even better.

The group subjected their newly grown diamonds to every possible chemical and physical test. The crystals withstood immersion in the strongest acids and they burned to yield only carbon dioxide; taken together, the tests provided convincing evidence that the crystals were pure carbon. X-ray diffraction revealed the characteristic pattern of the unique diamond atomic structure – a pattern indistinguishable from natural specimens. The synthetic crystals' density (3.53 grams per cubic centimeter), the distinctive optical properties, and extreme hardness also matched South African gems. There was no doubt whatsoever that the GE team had made diamond.

Even with such definitive proof, there was one crucial step left to be taken. A completely independent group of scientists – men who could not possibly have conspired with the Superpressure members to deceive the company – were brought in to duplicate the experiment. On January 18th and 19th, 1955, General Electric scientists Hugh Woodbury and Richard Oriani obtained their own independent sources of iron and graphite, and each made three runs with the belt apparatus on the four-hundred-ton hydraulic press. Hall and the other team members were not allowed to be present for this independent test, but Woodbury had completed one run under Hall's supervision and Hall provided additional detailed instructions. Novice operators Woodbury and Oriani shaped wonderstone gaskets, packed metal and graphite samples, ran the experiments, and analyzed their run products. They produced diamonds in all six experiments.

* * *

General Electric announced the creation of "Man-Made Diamonds," the new trade name for their product, at a well orchestrated press conference on February 15, 1955, two years to the day after ASEA's isolated, unpublicized success. Curious reporters were allowed to view a small pile of undistinguished-looking blackish grains under the microscope. The original research group posed for photographs, but was then sent away with strict instructions not to talk to anyone about the diamond-making process until the publicity died down. Bundy, Strong, Bovenkerk, and Wentorf went on a cross-country ski trip on frozen Lake George. In spite of the promotional efforts, not everyone was impressed at the display. Some reporters wondered why a company would invest a million dollars to produce what looked like a thimbleful of grit. Of what possible use could these tiny crystals be? In spite of such skepticism, the diamond world was stunned, and the world's diamond market was thrown into turmoil. At the time General Electric made its announcement, diamond trading, for both gemstones and industrial grade material, was controlled almost exclusively by the De Beers cartel. De Beers Consolidated Mines Ltd. of Kimberley, South Africa, had a virtual monopoly, selling what it wanted, to whom it wanted, at whatever price it wanted. To control the supply and price structure further, De Beers, it had long been rumored, controlled vast hoards of diamonds just waiting in vaults. Some suggested that for every diamond sold, one was put away for the future, and still their profit margin exceeded 50%. News of a potentially unlimited source of synthetic diamonds did not please the cartel.

De Beers stock dropped precipitously, while the total value of General Electric stock jumped overnight by more than $300 million. Some jewelers with sizeable investments in diamonds panicked, and bushels of mail arrived at Schenectady from around the world. Distraught gem merchants wanted to know if they had been ruined, outraged jewelers called GE's act unnatural, and would-be investors tried all sorts of ploys to obtain a piece of the pie. The value of stock and gemstone inventories returned to normal after a few days, but the diamond world had been changed forever.

Francis Bundy remembers from that turbulent time his favorite letter, one with a delightfully pragmatic approach to the GE announcement. Eight-year-old Chucky Singer of Peabody, Massachusetts, sent them a chunk of coal with a simple request. "I am sending you a piece of coal I found for you to make into a diamond in your machine. Please send it back to me." A few weeks later Chucky Singer received a small synthetic diamond in the mail.

Other correspondents were less optimistic. An anonymous California critic wrote: "We at the United States no you cant make real diamond for they are nature grown. You cant make gold no one can. They dig gold out of the ground and also diamonds. But no one can make them with a machine. That is just a lot of bull."

General Electric was happy to publicize its triumph, but the world wouldn't find out how it was done for several years. The original diamond-making team was given strict instructions about what to say and how to act. Bundy, Hall, Strong and Wentorf (in alphabetical order) were allowed to publish an article entitled "Man-Made Diamonds" in the prestigious weekly journal *Nature*, but other than mentioning high temperature and pressure, the piece gave no details at all about the process – a fact that infuriated some readers.[2] Scientific papers, the disgruntled subscribers argued, should include enough information for any reader to duplicate the results; the General Electric article was nothing more than a free advertisement.

General Electric rushed to submit several patents – one for the belt apparatus under Hall's name, others for aspects of the diamond-making process signed by various combinations of the original five team members. But when the documents reached the U.S. Patent Office they were immediately placed under a secrecy order; officials at the Commerce Department had not forgotten the fear of a diamond shortage during World War II. As a result of their ruling, no one else could see the patents – no one else could learn their secrets.

The General Electric management accepted this news with mixed feelings. Without international patent protection other companies might discover how to do it, too. But GE administrators also knew that as soon as the patents were made public, other researchers could take that hard-won information and attempt to improve the process. With the government-imposed secrecy, GE was given years to perfect the procedures without other researchers breathing down their necks.

One vital question remained for General Electric engineers to answer: would synthetic diamonds work as an effective abrasive? For several feverish weeks, from early January to mid-February, 1955, that question became the all-consuming passion of Hal Bovenkerk, who was to become a key player in the commercialization of synthetic diamonds. On December 16, 1954, the day after diamond was found, Bovenkerk was permanently transferred to the high-pressure group. Given the importance of diamond to GE's Carboloy operation, a liaison to the Detroit plant was needed. Bovenkerk, the only bachelor of the team, was the logical choice, and so began years of shuttling back and forth between New York and Michigan.

In early 1955 Hal Bovenkerk and Jim Cheney worked round the clock, completing dozens of fifteen-minute runs with the belt apparatus, sleeping near the presses. The belt had been refitted into a brand new table-top two-hundred-ton press, as the thousand-ton monster stood idly by. Each run produced a tough cylindrical plug of metal, graphite, and diamond somewhat smaller than a pencil's eraser. When cleaned in concentrated acids, each plug yielded a fraction of a carat of diamond. The machine shop was kept busy, too, with orders for a steady stream of wonderstone "flower pots" and replacements for the belt components that broke with annoying frequency.

Finally, after more than a hundred synthesis runs, they had what they needed – a precious stockpile of 23 carats of tiny, black synthetic diamond crystals. Bovenkerk hand-carried the supply to GE's Carboloy

Fig. 34 Harold Bovenkerk adjusting a belt apparatus, mounted in a thousand-ton press. (Courtesy of F. R. Boyd.)

plant in Detroit, where the grit was fabricated into a test grinding wheel.

The tiny crystals worked better than they could have imagined. Their sharp, tiny crystal faces proved ideal for grinding (fig. 34). Furthermore, the GE diamond makers found that they could carefully control the precise size of their crystals, thus providing diamond grit of unprecedented uniformity. A billion-dollar industry had been born.

* * *

You might think that the inventors of such a major process – a synthetic procedure that has yielded billions of dollars in sales – would have become wealthy men. But that is not GE's way. In exchange for their salary, researchers at General Electric sign away any rights to their patents. The four diamond makers, Bundy, Hall, Strong, and Wentorf, each received the standard company bonus – a twenty-five dollar savings bond for each patent. There were no instant promotions or extravagant increases in salary. In the view of management, the Superpressure scientists were just doing their job.

But with his key individual contributions so little recognized, and with so little financial reward for his years of work, Tracy Hall had had enough. "Saddened and hurt, I left General Electric, a company I had admired and aspired to work for since the age of nine," he wrote.[3] In April of 1955 Tracy Hall made up his mind to leave General Electric as soon as another job could be found. As news of diamond synthesis and Hall's special role in designing the belt spread, Tracy Hall became a hot property, and he had no trouble finding employment elsewhere. Philip Abelson, President of the Carnegie Institution of Washington, tried to get him to join the Geophysical Laboratory, where Joe Boyd and Hat Yoder had an active high-pressure program, but Hall wanted to return to his roots. In September 1955 he eagerly accepted the position as Professor of Chemistry and Director of Research at Brigham Young University in Provo, Utah. Though his starting salary was only $7500, significantly less than his GE income, he was treated well, given a large Quonset hut for a high-pressure laboratory and enough money to fill it with equipment.

Ironically, Tracy Hall was forbidden by company and government strictures to build or publish details of his own pioneering belt apparatus design. He made several trips to the U.S. Department of Commerce in Washington in an effort to alter the secrecy ruling, but to no avail. "The solution to my problem dawned one day when a man from the Commerce Department said, 'Hall, why don't you invent another apparatus?'" In a brilliant display of defiance and determination, Tracy Hall proceeded to design a completely different kind of high-pressure device of breathtaking novelty – and one that could duplicate the diamond-making feat.

With the help of a $10 000 grant from the Carnegie Institution of Washington, courtesy of Phil Abelson, and additional funding from the National Science Foundation, Hall invented a remarkable press, different from anything seen before. With the exception of von Platen's split sphere device, all previous device designs applied pressure between flat surfaces, along the single axis of a piston-cylinder or opposed anvil arrangement – an arrangement as familiar and logical as squashing a bug. Hall, in his second device, resorted to a quartet of pistons, one pointing straight down, the other three angling upward to meet at the faces of a tetrahedral sample assembly. The first "tetrahedral anvil press (fig. 35)," completed in 1957, provided a large sample volume, high pressures, and excellent access to the sample region.

Although the device was completely new, Hall could not be sure that the government would treat it as such. If, in their opinion, he had violated the secrecy agreement by building the machine, he would be

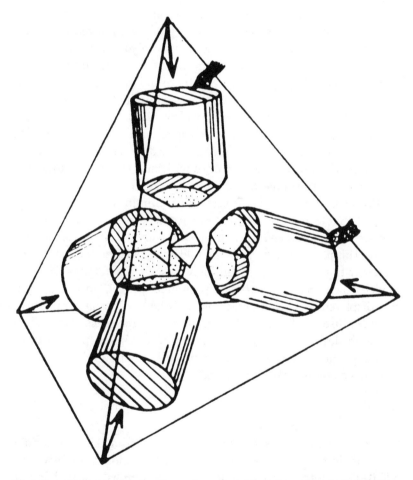

Fig. 35 Tracy Hall's tetrahedral-anvil press incorporated four pistons tipped with carbide anvils, each piston directed toward one face of a tetrahedral sample chamber. This device was Hall's second novel diamond-making design. (Courtesy of A. A. Giardini.)

subject to two years in jail and a $10 000 fine. "Never had I had so much anxiety and so many sleepless nights," he recalls. After much deliberation, Hall decided that his invention was clearly original and accordingly filed for a patent, submitted a descriptive article to *The Review of Scientific Instruments*, and gave a triumphant talk at the Spring 1958 meeting of the American Chemical Society.[4] He also sent out reprints of the article to dozens of scientists around the world.

The triumph turned to farce shortly thereafter when the Commerce Department official, who had encouraged Hall to try something new and promised to waive any restrictions, slapped a secrecy order on the tetrahedral press. It was too late to stop publication or retrieve all the reprints, but Hall was instructed to notify everyone who knew of the device that it was a secret and "conveyance of this secret to another was an act subject to the 2–year prison sentence and $10 000 fine." The fact that the experiment details were known to hundreds of scientists and the design published in an international journal did not seem to matter to government bureaucrats. Another trip to Washington by Hall failed to get the ban lifted, but Commerce officials did decide that Hall would have to notify only the hundred or so individuals who had either seen the press in person or had asked for reprints of the article.

Fortunately for Hall and the scientific community, the secrecy order was lifted a few months later, and by 1959 he was free to conduct high-pressure research and publicize his results without fear of reprisal. Hall received his patent, and he built and sold his tetrahedral apparatus commercially to more than a dozen high-pressure laboratories. His first model, which succeeded in making diamonds at his Utah laboratory, is still displayed as a central exhibit at Brigham Young University's Science Museum. The invention of the tetrahedral press placed Tracy Hall at the leading edge of high-pressure research and, perhaps more than any other factor, convinced many of his central role in General Electric's diamond making success.

* * *

Of all the world's scientists who heard the diamond-making news, none was more intrigued than Percy Bridgman. After more than forty years of failed attempts, he wanted to see the feat accomplished first-hand. In the summer of 1955, just a few months after the announcement,

Fig. 36 Percy Bridgman finally made diamonds during a visit to the General Electric Schenectady facility in March 1955. Bridgman (left) sits with fellow Nobelist Irving Langmuir, while (left to right) Robert Wentorf, Herbert Strong, and Francis Bundy look on. (Courtesy of H. M. Strong.)

Bridgman was asked by *Scientific American* to do a story on the historic breakthrough.[5] He paid a visit to The Knoll and had his chance. Bovenkerk, Bundy, Hall, Strong and Wentorf stood aside, partly in pride and partly in reverence, as the old master assembled an experiment himself and, at last, made diamonds (fig. 36). Francis Bundy described the scene: "We showed him all the details of the apparatus and the process, and invited him to do a 'diamond run' with his own hands. He did want to do it ... He put the cell together, put it in the high-pressure apparatus, did all the operations following our directions, and had a successful run. We gave him the diamonds he had made – and he treasured them."[6] (fig.37)

General Electric was happy to take advantage of the Nobel Prize winner's interest, and in late 1955 they asked him to become a consul-

Fig. 37 Herbert Strong, Tony Nerad, Percy Bridgman, and Harold Bovenkerk (left to right) discuss high-pressure diamond synthesis at GE, March, 1955. (Courtesy of H. M. Strong.)

tant to help with improvements in the process. After just two visits, however, he offered his resignation, noting that the GE diamond makers knew far more than he about their work – "Don't waste your money on me," he remarked.

* * *

By mid-1955 GE executives knew it was time to turn their years of research into a profitable business.[7] They did so with an old R&D addage clearly in mind: "For every hour of scientific research, there are 100 hours of development." The complex development stage of diamond manufacturing was transferred to the newly formed "Diamond Section" of the Detroit-based Carboloy Division, the single largest diamond consumer in the United States. The beginnings were extremely modest, with just a handful of two-hundred-ton presses, each equipped with a belt apparatus with one-half-inch-diameter sample chamber. These small devices only needed one-hundred-twenty tons to achieve diamond-synthesis pressures.

Carboloy engineer, J. Stokes Gillespie, was put in charge of the commercialization effort. Though a complete novice in the diamond-making game, Gillespie was a spark plug who got things done. He started with a single two-hundred-ton press and the first Detroit diamonds were made by August, 1955. By the end of 1955 a second press was added, and three more were placed on line in 1956.

The first to join Gillespie's team was Carboloy engineer, William Cordier, who had received his masters degree in mechanical engineering from Purdue. Cordier spent several summer weeks in Schenectady learning the diamond-making protocol, and then returned to Detroit with the first two-hundred-ton press. A short time later in December, 1955, Jim Cheney, who had assisted Herb Strong and other members of the original Schenectady team throughout the early 1950s, also transferred to Detroit, where he helped with the plant design and operation. Meanwhile, Hal Bovenkerk shuttled back and forth between the New York and Michigan facilities. In 1960 he, too, transferred full time to Detroit, eventually to become manager of research and development there.

Learning how to synthesize diamond grit was only the first step in the commercialization process. Successful marketing depended on production of an abrasive that was uniform in size, shape, and grinding ability. Rather than rush mediocre material to market, General Electric decided not to sell any diamonds at all until the process had been thoroughly tested and perfected. After hundreds of synthesis runs at various temperatures and pressures, the Detroit and Schenectady diamond makers discovered that they could control crystal shape with great precision.

Every crystal, diamond included, has a variety of possible crystal faces or forms. Diamonds in nature are sometimes found as tiny cubes with six square faces, other times as octahedra with eight triangular faces, and still other times with combinations of those and other more complicated forms.

Each surface in a diamond crystal has a different atomic structure; as the crystal grows, carbon atoms attach to some surfaces more readily than others. What may at first seem odd is that the face that grows fastest is the one that you *don't* see. It's something like trying to pile up marbles: it's easiest to add marbles to a layer parallel to the ground. You might start with a large flat area of marbles but, as you stack them higher and higher, that original plane – the easiest growth surface – disappears and the stack ends up in a pyramid with a point.

In exactly the same way, carbon atoms stack to build a diamond crystal. If the six cube faces grow fastest, the crystal will end up with six pointed corners and the eight-faced octahedral shape appears. Just the opposite happens if the eight octahedral faces grow fastest – you wind up with a cube with eight pointed corners and six faces.

Scientists don't know how to predict which forms will occur, but methodical research revealed that perfect octahedrons form most easily at higher temperatures close to the graphite-diamond transition – 1600°C at 60 000 atmospheres, for example. Cube-shaped crystals with just a hint of the octahedral forms predominate at lower temperatures, close to the melting point of the metal solvent – perhaps 1400°C at 60 000 atmospheres. At intermediate temperatures, elegant combinations of the cube and octahedron result.

Hal Bovenkerk and his coworkers in Detroit found that slow growth conditions produced crystals of remarkably uniform shape, excellent for rock sawing. Fast growth, on the other hand, resulted in a mass of more irregular crystals, ideal for grinding and polishing the toughest carbide parts. Imposition of temperature and pressure gradients in the synthesis chamber led to much more irregular shapes. They learned to grow needle-shaped diamonds, which are particularly useful in grinding wheel applications because they can be embedded deep in the wheel. These elongated diamonds invariably contain tiny inclusions of the solvent metal – iron, cobalt or nickel. These magnetic metals always crystallize along the needle axis, providing a simple and effective magnetic means of aligning the diamond shards in grinding tools.

In addition to searching for ways to improve the quality and quantity of the diamonds produced, the Detroit engineers had many routine chores. Every batch of graphite starting material was found to have different properties, and thus required slightly different optimal synthesis conditions. During 1957 and 1958 the Detroit group shared responsibility for determining those conditions. Eventually they learned that it was easier and cheaper to order enormous batches of graphite – enough for several years – to avoid the constant control experiments.

Another perennial problem was the carbide components of the belt apparatus, which GE made themselves at the Carboloy plant. Carbide pieces had to be replaced constantly. Every part failed eventually, perhaps after two or three hundred runs, but some components were apparently made from defective material and broke on the first try. The Detroit group devised tests to identify the best carbide pieces before the difficult and expensive grinding process. They checked the density, looked for microcracks and porosity with penetrating fluorescent dies, and performed ultrasonic examination to reveal internal defects. They also developed ways to "heal" defective carbide blocks, to reduce costs even more. These experiments weren't as glamorous as the original diamond synthesis, but they were a vital step in developing a practical, money-making process.

* * *

In November 1957, just a month after the Soviet Union launched its first Sputnik and shocked a technologically complacent United States to its core, General Electric began to sell "Man-Made Diamonds," its trade-marked synthetic diamond abrasive. Francis Bundy reflects the pride of the GE employees: "We like to think that we, by being first with Man-Made Diamond, took a little of the sting out of the Soviets being the first to put a satellite into Earth orbit." At the time of the announcement, General Electric's first diamond-making plant, the Diamond Section of the Detroit-based Carboloy Division, had manufactured just seven pints of diamond grit.

The General Electric sales force hoped eventually to be swamped with orders for the new product, which at last held the prospect of breaking the De Beers' monopoly. At first, however, with a limited range of abrasive products and a severe recession underway, the response was lackluster. Some buyers expressed reservations about patronizing GE, fearing a retaliation by the diamond cartel if the synthesis venture failed. It took General Electric several years and a lot of persuasive demonstrations to convince industry that Man-Made Diamonds performed as well, or even better, than natural material.

Ultimately, the balance was tipped by research, which provided General Electric's synthetic diamond with one clear advantage over De Beers' natural product. By carefully controlling the temperature, pressure, time, and starting materials of their runs, GE found it could produce a variety of abrasive materials with remarkably uniform properties. Synthetic diamonds could be tailor-made for specific applications: rock sawing and polishing, machining hard steels and carbides, gem cutting and polishing, and many other tasks.

Armed with an array of abrasives, General Electric undertook a massive publicity campaign to convince the machining industry to try the new products. Applications engineers in Detroit published numerous articles, repleat with reams of statistics and dramatic photomicrographs of controlled grinding and cutting experiments, in trade journals like *Carbide Engineering*, *Machinery*, *Grinding and Finishing*, and *American Machinist*. They produced spiffy brochures loaded with graphs and tables, all designed to convince potential buyers that Man-Made diamonds were superior to natural ones for manufacturers of automobiles, power tools, carbide components, and military hardware.[8]

It was a tough sell, but gradually the message got across. As early as mid-1959 *The Wall Street Journal* could report "GE already has weakened De Beers' hold on the U.S. industrial diamond market and forced the Johannesburg-headquartered company to offer special concessions to customers . . . GE's rise marks one of the more serious competitive threats to De Beers in its long history." In 1959 GE sold approximately three-quarters of a million carats of synthetic diamonds, capturing roughly 10% of the U.S. market, and they projected an increase to 3.5 million carats (about 1500 pounds) production the following year.

General Electric made steady inroads on De Beers' territory, with new products and steadily declining prices. In 1957 the price of synthetic diamond was twice that of De Beers ($5.30 per carat, compared to $2.87 for the natural material), but as production increased, the price dropped rapidly. By 1959 GE was able to match the price for natural industrial diamond abrasive – less than $3 per carat. The General Electric product was much more uniform in size and quality, however, and had better grinding qualities for many uses.

Right from the beginning the Norton Company, who had spent the better part of a decade trying to synthesize diamond, became GE's biggest customer. At first they paid a user subsidy to encourage the development of a domestic supply, and they continued to purchase GE product under special agreements. Increased production allowed GE to drop prices even further, to about $2.00 per carat by the mid-1960s.

* * *

The commercial implications of synthetic diamonds were, understandably, foremost in the minds of GE management, but Francis Bundy, Herb Strong, and Bob Wentorf were focused on the years of fascinating science in front of them. In 1955 General Electric eagerly encouraged this research, and the years following the discovery saw a flurry of experiments on diamond stability, effects of different temperatures and pressures, substitutes for iron as a catalyst, and improved pressure-generating components. Two spanking new thousand-ton presses were added to the Schenectady research operation and each was outfitted with a carbide belt apparatus – the device that had proven most efficient for diamond making.

After hundreds of experiments, the scientists discovered that virtually any combination of the metals iron, nickel, chromium, cobalt, manganese, tantalum, ruthenium, rhodium, and platinum could be used to induce diamond growth. The key, it turned out, was to reach temperature and pressure conditions at which diamond was the stable form of carbon *at the same time* that the metal was in a liquid state. The GE scientists discovered that the liquid metal not only dissolved the carbon, thus providing a steady supply of isolated carbon atoms, but also catalyzed the diamond crystal growth.

Given their unique diamond-making ability, the General Electric researchers were in an ideal position to study many of the longstanding questions about carbon. They were especially eager to learn the range of temperature and pressure under which diamond forms, but they needed a high-pressure device that would extend well beyond the belt's 60000 atmospheres and 2000°C. They also needed to devise a means of calibrating temperatures and pressures at those extreme conditions. Francis Bundy and coworkers developed a modified "superbelt" apparatus with thinner, more pointed anvils that allowed them to reach 150000 atmospheres and attempt brief temperature excursions to 5000°C – conditions high enough to melt carbon (fig. 38).

Bundy undertook a detailed study of the phase boundaries among graphite, diamond, and carbon liquid.[9] Bundy established the range of conditions under which graphite converted to diamond, and he discovered the carbon "triple point" – the unique combination of temperature and pressure (about 4100°C at 125000 atmospheres) at which graphite, diamond and liquid coexist (fig. 39). The extreme conditions in the superbelt apparatus also enabled Bundy to convert graphite directly into diamond without using any metal solvent or catalyst. Graphite, which Percy Bridgman had dubbed "nature's best spring," had at last been pushed to its breaking point.

In the years following 1955, GE was under pressure to fend off the competition. In particular, they had to be sure that all the original

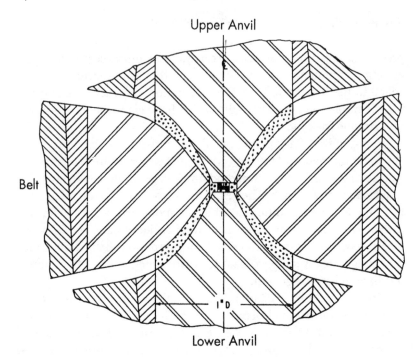

Fig. 38 The highly tapered pistons of the superbelt enabled Francis Bundy to reach the extreme temperatures and pressures at which graphite could finally be converted directly to diamond without a metal solvent. (Courtesy of F. R. Bundy.)

diamond patents were indisputably correct. The first General Electric patent claimed that any substance rich in carbon, not just graphite, could be converted to diamond with the GE process. That was a tall claim, open to challenge by other companies unless a variety of carbonaceous materials could be shown to work. Bob Wentorf volunteered for the job, and began work on one of the most bizarre and amusing series of experiments in the history of high pressure.

Carbon-rich materials are everywhere: plastics, sugar, wood, paper, glue, and all living things are loaded with the element carbon. Most scientists might have run diamond-making experiments on fancy-sounding carbon-based chemicals like polyethylene or pentanedione, but Wentorf, true to his whimsical nature, turned to more familiar substances. In high-pressure runs of remarkable flamboyance he squeezed "Black Diamond Roofing Tar," aged maple wood, moth flakes, and his favorite brand of crunchy peanut butter. The latter experiment was a

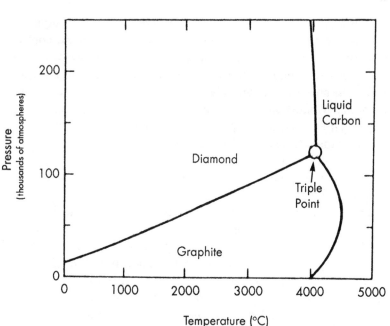

Fig. 39 Francis Bundy determined the carbon triple point – the unique combination of pressure and temperature at which graphite, diamond, and liquid carbon coexist – to be approximately 4100°C and 125 000 atmospheres. (Courtesy of R. DeVries.)

consequence of Wentorf's inclination as "an avid snacker." He always kept a box of crackers and jar of peanut butter in his right-hand desk drawer. "The peanut butter turned into tiny green diamond crystals," Bob laughs. "It was because of all the nitrogen." The novel peanut butter experiments got Wentorf more press coverage than just about anything else the GE diamond makers did,[10] but those runs also made a telling point. General Electric *could* turn just about any carbonaceous material into diamond.

After Wentorf repeated the famous peanut butter story for what must have been the thousandth time, Herb Strong gave the punch line: "The pity is you can't turn diamonds into peanut butter." It's a well rehearsed line, but there is little doubt that, if given the chance to invent just one product in his life, Bob Wentorf would have been just as happy to have discovered peanut butter.

Wentorf's studies on diamond and graphite led him to investigate other compounds as well. One of the greatest lures of high-pressure research is the prospect of discovering entirely new materials with unusual properties. In 1956, Bob Wentorf made such an extraordinary discovery – a completely new abrasive that in some ways works even better than diamond. Wentorf studied the compound boron nitride, a soft white powder that normally has a layered crystal structure just like graphite. Wentorf suspected that boron nitride, like graphite, might have a hard, abrasive high-pressure form. His successful synthesis of "cubic boron nitride" provided the world with another superabrasive.[11]

Cubic boron nitride, trademarked as "Borazon," turned out to be almost as hard as diamond and every bit as resistant to attack by acids and other chemicals. Unlike diamond, it is stable to more than 1500°C in air, a temperature at which diamond has long since burned away. But the most important use of Borazon is found in tools for the steel industry. Diamond, heated by the grinding process, reacts chemically with iron-bearing alloys and is ineffective. Borazon, on the other hand, cuts through iron-bearing alloys quickly and easily. Given these unique characteristics, a variety of cubic boron nitride products were added to General Electric's catalog of synthetic abrasives in 1969.

Most scientists, particularly Percy Bridgman, were delighted to hear of this novel high-pressure phase. Eric Lundblad and his ASEA colleagues, however, were chagrined by the news. They went back to earlier high-pressure experiments they had conducted on the same compound and, sure enough, they'd made cubic boron nitride themselves without knowing it.

* * *

The year 1959 was pivotal in the commercialization of synthetic diamond. Four years earlier GE had grudgingly accepted the secrecy order, knowing that it granted them extra time to study the process and an opportunity to develop a system that could monopolize the manufacturing of synthetic diamond. But they were constantly aware that all their efforts could be undercut by foreign rivals. If another group – especially De Beers – discovered how to make diamond before GE was allowed to file for international patent protection, then the American company would lose its foreign rights over the process.

General Electric began selling its synthetic diamonds in 1957, and United States industry was at last assured of a secure supply of diamonds. But commercialization carried certain risks: every synthetic diamond contained telltale traces of the secret metal solvent used to create it, and by 1958 everyone in the business knew that iron and nickel had something to do with GE's secret process. Foreign research groups were rumored to be close to cracking the diamond-making riddle, and with the U.S. prohibition against filing foreign patents still in place, General Electric stood to lose its lucrative potential world market.

* * *

The sense of urgency was heightened when it was learned that an American group had independently accomplished the diamond-making feat.[12] Armando A. Giardini, a high-pressure specialist at the U.S. Army's Electronics Research and Development Laboratory at Fort Monmouth, New Jersey, first became interested in the diamond problem as a graduate student at the University of Michigan in 1953. Giardini and classmate Richard S. Mitchell were employed by Professor Chester B. Slawson as research assistants on a grant from the Office of Naval Research. Slawson believed that diamond could be made by the decomposition of mercuric carbide under high pressure and temperature. As early as 1951 and 1952 the Michigan group saw weak but inconsistent x-ray diffraction signatures of diamond. Their chief problem was the instability of the mercury starting material. "More often than not, the stuff exploded as we increased the pressure," Giardini recalls. The inconclusive results were written up by Slawson for the January, 1953, issue of *The American Mineralogist*.[13] "Since we could not consistently reproduce our results . . . he made no mention of diamond having been synthesized. This is the way science used to be," he adds wistfully.

Giardini spent the next several years working for the Carborundum Company before returning to Michigan for a Ph.D., earned in 1957. From Michigan he went to the Army lab in New Jersey, where he convinced the manager, Dr. Samuel B. Levin, to support a diamond-making project. With his coworker, Lt. John E. Tydings, Giardini reviewed all the current designs for high-pressure apparatus (fig. 40). First, they read all of Bridgman's articles and visited his Harvard laboratory late in 1957. They also met with Joe Boyd and Hat Yoder at the Geophysical Laboratory and they travelled to Schenectady to talk with the General Electric scientists. "The latter were very cordial, but totally secretive. . . . All their equipment was covered with paper drapes during our visit."

Fig. 40 Armando Giardini (kneeling) and John Tydings, *circa* 1959, at the U.S. Army laboratory at Fort Monmouth, New Jersey, where they synthesized diamonds prior to the patent disclosures of General Electric. (Courtesy of A. A. Giardini.)

Giardini and Tydings spent a year and a half essentially reinventing the wheel. Their independent creation, the "Supported Stepped Piston-Cylinder Apparatus," was a device similar to the belt in capabilities, as well as concept. They, too, used opposed tapered carbide pistons, a carbide girdle reinforced with steel binding rings, and confining gaskets of pyrophyllite. They, too, could reach temperatures and pressures close to 100 000 atmospheres at high temperatures.

With their powerful device they went straight to work on making diamond. The traces of nickel in GE's commercial synthetic diamond provided a critical clue; they, too, heated and squeezed graphite with nickel. "We couldn't seem to do it. We had noticed everything from fused graphite spheres to lumpy masses of metal units in our reaction products" – but no diamonds. The solution to their problem came at the suggestion of Alvin Van Valkenburg, a high-pressure scientist from the National Bureau of Standards who was visiting the New Jersey facility. "He asked if we had determined what was the cause of the lumps in the metallic stubs. We hadn't, not once. . . . While he was there we placed the metal in some aqua regia [acid], and there before our eyes were revealed beautiful crystals of diamonds. I've never spoken much of this; I still feel foolish, and still feel grateful to Van." Giardini, Tydings and Levin quickly prepared a manuscript and submitted it to *The American Mineralogist.*

As early as June 1959, rumors were circulating among high-pressure workers regarding the imminent declassification of GE designs. The government asked expert witnesses, including Alvin Van Valkenburg, to examine the original General Electric patents, as well as GE's strongly worded petition for the lifting of secrecy. Van Valkenburg recounted vivid memories of his visit to the patent office. After a careful review of the documents Van Valkenburg realized that Giardini's soon-to-be-published results paralleled the General Electric claims. There was nothing vital to be kept secret. He and other experts strongly supported the end to restrictions on the information, and within a few weeks, by the end of September, 1959, the secrecy order was lifted. The very next day lawyers around the world filed the ready and waiting patents on General Electric's behalf. The filings came none too soon, for a week later De Beers submitted its own diamond-making patents. The South African company was dismayed to find they were just days too late.

After four years of waiting, the original team of diamond makers was at last free to publish the details of their breakthrough. "Preparation of Diamond" by Bovenkerk, Bundy, Hall, Strong and Wentorf (again in alphabetical order) appeared in the October 10, 1959, issue of *Nature*,[14] and the information was repeated in the news columns of *Science* and other magazines. A few months later *The Review of Scientific Instruments* for February, 1960, contained Tracy Hall's detailed description of the belt apparatus.[15]

At last the world had learned the secret. Now, anyone with a press could become a diamond maker.

Notes

1. C. Guy Suits, "Scientific competition," in *Speaking of Research* (NY: Wiley, 1965), p. 377.

2. Francis Bundy, H. Tracy Hall, Herbert Strong, and Robert Wentorf, Jr., "Man-made diamonds." *Nature* **176**, 51–55 (1955).

3. H. Tracy Hall, "The transformation of graphite into diamond." *Newsletter of the American Association for Crystal Growth* **16**, 2–4 (1986), p. 4.

4. H. Tracy Hall, "Some high-pressure, high-temperature apparatus design considerations: equipment for use at 100,000 atmospheres and 3000°C." *Review of Scientific Instruments* **29**, 267–275 (1958); see also Judith Milledge, "Natural and synthetic diamonds." *The Times Science Review*, Spring 1960, 8–13.

5. Percy Bridgman, "Synthetic diamonds." *Scientific American* **193**, 42–46 (1955).

6. Francis Bundy, "Bridgman's own struggle with the physics revolution." *Physics Today* May 1991, p. 63.

7. Many details of the early commercialization of synthetic diamonds were provided during interviews and correspondence with Harold Boverenkerk and William Cordier.

8. A typical pamphlet is "Newest tool for Industry! General Electric Man-Made Industrial Diamonds," Detroit, Michigan: General Electric Company, 1959, 12 p.

9. Francis Bundy, "Direct conversion of graphite to diamond in static pressure apparatus." *Journal of Chemical Physics* **38**, 631–643 (1963); see also Francis Bundy, "Diamond synthesis with Bridgman opposed-anvil apparatus." *Science* **146**, 1673–1674 (1964).

10. A press release was issued by GE's Mechanical Investigations Section: Robert Wentorf, Jr., "Diamond directly from Arachis hypogaea," GE News Item No. C-58–63, September 30, 1963.

11. Robert Wentorf, Jr., "Cubic form of boron nitride." *Journal of Chemical Physics* **26**, 956 (1957); Francis Bundy and Robert Wentorf, Jr., "Direct transformation of hexagonal boron nitride to denser forms." *Journal of Chemical Physics* **38**, 1144–1149 (1961).

12. Armando Giardini provided details of the Army high-pressure research program research effort through interviews and correspondence. See also: Col. Raymond H. Bates, "Chapter II – summary of progress," in: *U. S. Army Institute for Exploratory*

Research 1958–1961. (Fort Monmouth, NY: U.S. Army Signal Research and Development Laboratory, 1962), pp. 4–19.

13. A.A. Giardini, J.E. Tydings, and S.B. Levin, "A very high pressure-high temperature research apparatus and the synthesis of diamond." *American Mineralogist* 45, 217–221; A. Giardini and J.E. Tydings, "Diamond synthesis: observations on the mechanism of formation." *American Mineralogist* 47, 1393–1421 (1962).

14. Harold Bovenkerk, Francis Bundy, H. Tracy Hall, Herbert Strong, and Robert Wentorf, Jr., "Preparation of diamond." *Nature* 184, 1094–1098 (1959).

15. H. Tracy Hall, "Some high-pressure, high-temperature apparatus design considerations," op. cit.

CHAPTER 9

RISKY BUSINESS

Superman's challenge is to save Lois Lane and Jimmy Olsen from revenge-hungry tribesmen, who have lost the gemstone eye of their idol-god. Unscrupulous men have stolen and then misplaced the enormous jewel, but ever resourceful Superman spies a lump of coal, squeezes it with his super powers while baking it with his x-ray vision, and produces a sparkling diamond replica of the lost eye. The warriors, poised to dispatch Lois and Jimmy, are suddenly appeased and all is well.

IT SEEMS EXTRAORDINARY THAT when this TV episode first aired in 1957, just a couple of years after General Electric's diamond-making breakthrough, the process had already become a part of popular culture. So great was the publicity surrounding the technological breakthrough, that most Americans knew you could make diamonds by heating and squeezing coal.

In 1959, when the world learned the details of how to synthesize diamonds, the process went beyond popular culture – it became part of the scientific record. General Electric scientists had told their peers exactly what to do to make diamonds, and anyone with a decent high-pressure laboratory could repeat the process. Naturally, a lot of them tried, and by Hal Bovenkerk's estimation more than two dozen groups did it within a year.

Armando Giardini and John Tydings, the scientists whose diamond-making efforts at Fort Monmouth, New Jersey, had succeeded in mid-1959 just prior to GE's revelations, were in the best position to duplicate the General Electric recipe. It was a straightforward matter for Giardini and Tydings to crank their device up to 85000 atmospheres and 1460°C with a graphite and nickel sample, and they produced diamonds on their first try.[1]

* * *

Francis R. "Joe" Boyd, a high-pressure specialist at the Carnegie Institution of Washington's Geophysical Lab, tackled diamond making as soon as he heard about the GE process.[2] As an undergraduate at Harvard University in the 1940s, Boyd entered geology because he was captivated by one brilliant teacher, Kirtley Mather, who brought a drama and richness to the science of rocks. Mather conveyed the poetry of geology, which grapples with epic global forces, incomprehensible spans of time, and the origins of life itself. By the summer of his senior year, Joe Boyd had earned the chance to be field assistant to Harvard's maverick junior faculty member and high-pressure pioneer, George Kennedy.

Boyd eventually became a Kennedy graduate student at Harvard, where he combined field work and high-pressure experiments in a Ph.D. thesis on the volcanic rocks of Yellowstone National Park. He graduated as a top science draft pick for 1953, and he went straight to the Carnegie Institution of Washington's Geophysical Laboratory, where he has been a research scientist ever since.

Joe Boyd has a well-earned reputation for getting things done. In 1977 he organized an epic field trip for 100 geologists from around the world who had converged on the American Southwest to study kimberlites – the rocks that hold diamonds. The entire group was to take a raft trip down the San Juan River to reach a key kimberlite rock exposure, but an unusually dry period had left the river too low for rafting. No problem. Boyd placed a call to his friend, Vince McKelvey, Director of the United States Geological Survey. McKelvey saw to it that the Navaho Dam, upstream of the raft party, was opened long enough to supply the necessary water.

Boyd made his mark in high-pressure research by teaming up with Joseph L. England, who is remembered by his many friends as an amiable, low-key fellow (fig. 41). England earned a bachelor's degree in physics and, like his father before, was a skilled machinist. He never developed his own research program, but was happy to work closely on Joe Boyd's projects. Together they set about constructing a novel two-stage piston-cylinder apparatus. The device, a clever stepped piston-cylinder, was one of the first to achieve high temperatures at 100 000 atmospheres. Boyd and England used the cumbersome press to conduct *tour de force* studies of rock and mineral melting at deep crust conditions. They wanted to know how igneous rocks – rocks that have melted and recrystallized – form deep within the earth.

Together Boyd and England built an impressive laboratory, with four piston-cylinder devices in steady operation, with which they tackled classic questions about the origins of rocks. While they could reach

Fig. 41 Joe England (foreground) and F. R. Boyd made diamonds at the Carnegie Institution of Washington's Geophysical Laboratory in January 1960, shortly after General Electric released details of the process. (Courtesy of F. R. Boyd.)

diamond-making conditions with their unique facility, they didn't spend a lot of time working at those extremes. There was so much to learn in the easier, safer 10000– to 50000–atmosphere range of routine operation. But as soon as they heard about the General Electric process, they repeated the diamond-making feat with ease, squeezing graphite and nickel to 75000 atmospheres.

"We did it for sport," Boyd recalls of the diamond-making runs completed just a month or so after the GE announcement. "It was really an exciting experiment to make." The Carnegie Institution thought so too, and they issued their own press release on January 22, 1960, describing the tiny, faceted black crystals.[3] Their one-page statement emphasized the role of the nickel catalyst, and concluded with an intriguing idea: "Since rocks containing diamonds do not contain an uncombined metal, natural diamonds cannot have formed by [the GE] process. Further studies may reveal more about the growth of natural diamonds as well as the conditions of formation of the rocks in which they are found."

Boyd and England repeated the diamond trick a few times, but never with any commercial goals in mind. They simply wrote a short note on "Synthesis of Diamond" as part of their longer report on mantle mineralogy in the Carnegie Institution of Washington's *Year Book* for the 1959–60 academic year.[4] Boyd heard a few questioning remarks from Carnegie Institution trustees, who wondered why the Geophysical Lab hadn't tried to make diamonds first, but that wasn't why he did high-pressure research, and the Carnegie Institution wasn't in the patent business.

One unforeseen repercussion of Joe Boyd's success was a noticeable cooling of relations with his former mentor, George Kennedy. "He never got over my being the first [of the two men] to make diamonds," Boyd speculates.

* * *

After the diamond-making procedure had become routine, the General Electric research team set its sights on larger crystals. Was it possible to grow one-carat gemstones? The resulting research program produced not only spectacular diamonds, but also one of the most spectacular accidents in the history of diamond making.[5]

Explosions are a fact of life in high-pressure research; the bigger the press, the bigger the blowout. The catastrophic GE blowout occurred as the result of what is known as a Langmuir experiment.

In the 1920s Irving Langmuir, a Nobel Prize winning General Electric research scientist, had taken on the challenge of creating a longer-lasting light bulb. At that time, light bulbs failed frequently because air would leak into the bulb, burning up the hot filament. Langmuir wondered whether the problem could be avoided by getting rid of the vacuum and using a small amount of an inert gas like argon in the bulb instead. Conventional wisdom scoffed at the idea: scientists argued that heat from the glowing tungsten filament would flow right to the glass via argon atoms, making an impossibly hot and inefficient bulb. Langmuir tried the experiment, but he started the easy way. Instead of constructing a vacuum bulb with just a little argon, he used a full atmosphere of argon – a full 14.7 pounds per square inch. If a little argon caused too much heating, he reasoned, then a lot of argon would exaggerate the effect and make it much easier to measure. The result surprised everyone – even with a full atmosphere of argon the bulb worked fine, with a minimum of heating. Since that time, GE scientists referred to any experiment that exaggerates the variable of interest as a Langmuir experiment.

The diamond team's ill-fated Langmuir experiment was born when Herb Strong wondered why synthetic diamond crystals were always so small – usually no more than a few hundredths of an inch across. Would the addition of another element stimulate crystal growth? Following a suggestion by high-pressure expert George Kennedy, who had then moved from Harvard to UCLA, Strong decided to study the effects of hydrogen on diamond growth. In the spirit of a Langmuir experiment, Strong decided to add not just a little hydrogen, but a huge amount in the form of a chunk of polyethylene plastic, which averages two hydrogen atoms for every one of carbon. At conditions of diamond formation, Strong was to learn, polyethylene breaks down to form carbon and hydrogen – lots of hydrogen. The trouble is that hydrogen at high pressure is one of the world's best wedges, exploiting every possible crack and defect in metal and gasket components. The hydrogen atoms penetrate deeply, weakening the structure as they go. To make matters worse, the hydrogen is chemically reactive, combining with iron to form weak, brittle iron hydride, which is highly susceptible to failure at high pressure.

The belt apparatus blew up during Strong's very first attempt. With a deafening detonation and impact that shook the entire knoll research complex, the hydrogen gas escaped and then ignited in a miniature re-enactment of the *Hindenburg* disaster. Shrapnel sprayed out across the room and ricocheted off the walls and ceiling. Luckily, no one was injured in the explosion, and thereafter, anyone at GE who tried to study polyethylene at high pressure and temperature proceeded with great care. The team would have to find another way to make large synthetic gems.

* * *

After a decade of sporadic effort and much trial and error, Bob Wentorf and Herb Strong learned how to grow exceptional diamond crystals, more perfect than any in nature, at the rate of about one carat a week.[6] Most GE diamond production incorporates a sample of graphite and molten metal at constant temperature: diamond grows as carbon dissolves. It is difficult to control the exact concentration of carbon throughout the sample chamber, so numerous diamond crystals start growing at the same time and none of these crystals gets very large (fig. 42).

An important clue to obtaining big crystals came from Jim Cheney, who noticed that larger crystals sometimes formed at the cooler ends of some experiments with a temperature gradient across the sample chamber. In 1960 Bob Wentorf realized that more carbon dissolves in hotter metal, so a temperature gradient (which is relatively easy to control) provides an ideal way to control carbon concentration. Wentorf discovered that by putting diamond powder at the hotter end of his sample chamber and a diamond seed crystal at the cooler end, the seed diamond grows at the expense of the diamond powder.

This technique worked sometimes, but the seed crystal had an annoying habit of floating to the hotter end of the sample. Herb Strong, working closely with technician Roy E. Tufts, came up with a clever fix by partially embedding the diamond seed in ordinary salt – a solution that led to the routine production of beautiful one-carat crystals (fig. 43). Strong and Wentorf learned to impose exquisite control of the high-pressure, high-temperature environment in the growth of large diamonds. They were able to maintain a precise temperature gradient of about 10°C across the sample chamber for many days, as carbon atoms migrated through a molten metal bath and, one-by-one, attached themselves to the growing diamond crystal.

Fig. 42 Flawless diamond crystals more than one carat in weight were grown by GE scientists in the late 1960s. Herbert Strong (left) and Robert Wentorf, Jr. (right) show a collection of synthetic gems to GE vice president of research and development Arthur M. Bueche in May 1970. (Courtesy of H. M. Strong.)

Wentorf and Strong found that growth rates increase with larger temperature gradients, but the resultant crystals displayed more defects and inclusions. After a decade of sporadic effort, the GE researchers learned the tricks to produce magnificent single-crystal diamonds, approaching two carats in size. Under normal circumstances the GE crystals were yellow, the result of ubiquitous nitrogen impurities, but the addition of a nitrogen-binding element like aluminum or titanium facilitated production of purer, colorless gems. The General Electric scientists also used boron-rich starting materials to grow remarkable blue diamonds, which are extremely rare in nature.

Experts doubt that large synthetic diamond crystals will ever make a dent in the gem market; they take too much time to grow, the giant presses are too expensive to run, and synthetic gems lack the glamor of stones formed a hundred miles deep in the earth. But scientists and engineers covet these glorious synthetic stones for another reason: in several respects synthetic diamonds can be even better than natural dia-

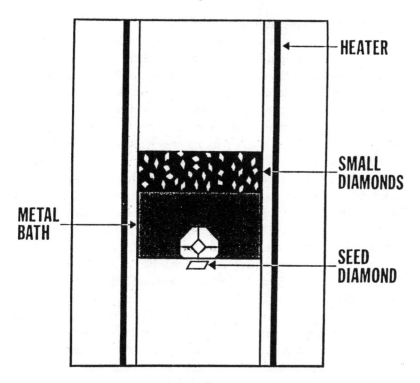

Fig. 43 A schematic diagram of a high-pressure assembly reveals how single-crystal diamonds, starting with a seed diamond at the cooler end of the assembly, are grown at the expense of small diamonds in the hotter end. (Courtesy of Robert Wentorf.)

monds. All carbon atoms contain exactly six protons in their nucleus, but the number of neutrons – the other heavy nuclear particle – varies from atom to atom. The commonest carbon isotope, carbon-12, has six neutrons, but about one of every hundred carbon atoms has seven neutrons, to make carbon-13. Natural diamonds incorporate this distribution, but GE scientists learned to synthesize isotopically pure carbon-12 or carbon-13 crystals.[7]

Using nuclear age technology that separates atoms according to their individual isotopes, researchers at Oak Ridge National Laboratory and other facilities purify carbon-12 and carbon-13. It's an expensive proposition, but you can now buy these carbon isotopes off the shelf from government labs. To make isotopically pure diamond crystals, scientists usually begin with an isotopically pure carbon in the form of graphite and synthesize diamond grit in the usual fashion, with high temperature, high pressure, and a metal solvent. This isotopically pure diamond powder can then be used to grow large diamond crystals in which every carbon atom is exactly like every other.

Most of the properties of diamond are unaffected by such uniformity. Carbon-12 and -13 diamonds produced at GE's Worthington, Ohio, plant are hard, transparent, and beautiful just like other diamonds. However, one key feature – the diamond's ability to conduct heat – seems to increase enormously in isotopically pure crystals. Heat moves by wiggling atoms. When different weight atoms like carbon-12 and carbon-13 occur together, the heat flow is slightly disrupted by the different vibration speeds of the two isotopes (heavier isotopes vibrate more slowly). But when all the atoms are exactly the same, heat is able to travel in steady waves. Thus, it turns out that synthetic carbon-13 diamonds, with 99.9% or more isotopic purity, are the best known heat conductors in the world.[8] These crystals now play a critical role in sensitive electronic gear for high-speed computers and military applications, where cool circuits work better than hot ones.

Another intriguing attribute of isotopically pure carbon-13 diamonds is that they are denser than natural diamonds, which are predominantly carbon-12. The greater density is due primarily to the fact that carbon-13 diamonds average almost one more neutron per atom than normal diamonds. In addition, because of more subtle interactions between each atom's nucleus and its electrons, the carbon–carbon spacing in carbon-13 diamonds is ever so slightly smaller than in carbon-12 diamonds. As a consequence, carbon-13 diamonds contain more atoms per unit volume than any other known material at room pressure.[9] In 1993, a team of scientists at Purdue University in Indiana and Argonne National Laboratory in Illinois, working with GE scientists, found that such synthetic "dense diamonds" are actually harder than natural stones. Carbon-13 diamonds are thus the hardest known material.[10]

* * *

As the quest for large single-crystals continued, efforts to develop the exact opposite – extremely fine-grained diamond grit – also gained momentum. Some natural diamonds occur not as isolated crystals, but as hard black masses of tiny intergrown crystallites. These forms of diamond, called carbonado and ballas, feature grains that are locked together with diamond-to-diamond bonds, producing a solid that can actually be tougher than single-crystal diamond. The diamond makers realized that while tiny crystals embedded in a saw blade or abrasive wheel work fine for ordinary cutting, grinding, and polishing, solid diamond tools would be much better for more demanding operations. Single-crystal diamonds are too expensive and break too easily along certain crystal planes for use as tools, but large chunks of carbonado are ideal, and many scientists were confident that they could duplicate nature's carbonado process.

Bob Wentorf and his Schenectady colleague William A. Rocco tackled the problem in 1970, and within a year they had found a solution.[11] Using tried-and-true diamond-making protocol, Wentorf and Rocco subjected a mixture of cobalt metal and extremely fine-grained synthetic diamonds to the same temperatures and pressures used in diamond making – roughly 60 000 atmospheres and about 1400°C. At those conditions, the tiny diamond crystals deform and regrow into a tightly bonded mass of more than 85% diamond, with some interspersed cobalt. The cobalt can then be dissolved away in acid, leaving a diamond-hard aggregate that resists fracturing.

Detroit engineers turned Wentorf and Rocco's laboratory procedures into a valuable new product which they called "Compax." Compax tools and wire dies begin as solid black masses up to several inches across. The diamond is then cut and shaped with an electric spark technique that vaporizes the carbon atom by atom. These tools have hundreds of uses, but sintered diamond plays an especially important role in rock drilling, which uses drill bits studded with protruding diamond cutting teeth. These drill bits have to cut through thousands of feet of solid rock, and every pause to pull up a mile or more of drill rig and replace the drill bit costs thousands of dollars. So long-lived diamond drills save companies (and, one hopes, consumers) a fortune.

With discovery after discovery to its credit, the General Electric high-pressure research laboratory at the Knolls had become famous as "The Diamond Mine." Many prominent high-pressure scientists, as well as astronaut Harrison "Jack" Schmidt, actor Ronald Reagan, and inventor Buckminster Fuller, were among the dignitaries who came to tour the facility.

Throughout the 1960s General Electric's Detroit diamond-making facility also grew. With a rapidly increasing demand for its man-made diamonds, GE quickly added ranks of thousand-ton presses with larger-diameter belts. More personnel were constantly needed to man the new

Fig. 44 General Electric employees Anthony Nerad, Jack Kennedy, A. L. Marshall, and a testing technician (left to right) pose at the GE Carboloy Department's Diamond Section in Detroit, Michigan, *circa* 1957. (Courtesy of Francis Bundy.)

equipment and devise new techniques for machining with diamond and cubic boron nitride. It was obvious that they would soon expand beyond the limits of the Detroit site (fig. 44).

* * *

The move to the new facility took place in 1968, when all abrasive production was transferred to Worthington, Ohio. The General Electric pioneers who saw the first glimmer of synthetic diamond in 1954 could scarcely have imagined that modern diamond factory.[12] The sprawling plant, graced by green lawns and a shallow sculpted pool, lies near the northern-most point of the Columbus beltway, just a couple of miles from Francis Bundy's birthplace. The low, flat-roofed tan buildings that house the diamond factory cover acres of central Ohio plains.

For years the purpose of the plant was shrouded in secrecy. Known only as the "GE Specialty Materials Department," the facility's mission was revealed to relatively few people. Even Ohio's Governor and legislators were unaware that Columbus had become the diamond capital of the world.

Today, diamond making is still big business for GE. A few years ago, in a renewed effort to publicize their product and expand markets, the General Electric diamond makers assumed a much higher profile. They changed their name and, for a brief time before corporate management became nervous, invited scientists and industrial representatives to tour the factory. Other than the modest "GE Superabrasives" sign out front, however, there is nothing to suggest the spectacular product created inside.

Behind the security guards and oversized doors lie the heart and soul of GE Superabrasives: the vast rooms where tons of diamonds are grown. Hundreds of workers make their living here, operating row after row of massive presses, each capable of producing pounds of diamonds a day. One room the size of an airplane hanger contains dozens of thousand-ton presses, each standing twice the height of a man, each capable of generating more than 60 000 atmospheres of pressure. An incessant hum of hydraulic devices fills the clean, brightly lit room, with its tan cinder block walls and machines painted pleasing shades of blue and yellow (the yellow parts are the ones that move, in accordance to government regulations).

All the presses are equipped with belts remarkably similar in form to Tracy Hall's original. Ranks upon ranks of thousand-ton presses, perhaps a hundred in all, accommodate two-foot diameter belts, while four-foot diameter belts fit into several much larger presses. The factory also has a few monster presses that are two stories tall with awesome five-foot diameter belts. Open metal stairs lead down to the thick concrete foundations on which the largest presses and hydraulic apparatus must rest. Bold warning signs admonish workers to wear ear protection before descending into the deafening pit, where noisy compressors and blowers throb continuously.

Every aspect of the spacious facility is designed for efficiency. The carbide components of the belts are tough, but they do wear out, and once in a while a press will "blow out" with a percussion that rocks the whole factory. With so many presses in constant use, breakage is a routine occurrence, so the factory walls are lined with sturdy metal shelving, stacked with brand new anvils and belts just waiting for quick installation, and the factory ceilings are criss-crossed with steel I-beams and heavy-duty cranes to aid in repairs.

The thousand-ton presses are astonishingly fast and efficient. In each diamond-making device, as in Tracy Hall's original, two gracefully tapered pistons squeeze a carbon-rich sample that has been packed into the hole of the doughnut-shaped belt. It takes a few minutes to load a sample into the hole and a few minutes more to apply heat and pressure. The Worthington belts have been equipped with a clever system of auxiliary sample holding rings to double their productivity. The two rings connect at a pivot point, like the two circles of a figure eight. The belt operator, who wears a blue uniform, protective glasses, and thick gloves, swings one fully loaded belt into position so that the ring of steel and carbide is perfectly aligned, ready to be squeezed between the carbide pistons and heated by a powerful surge of electrical current. As the loaded belt rotates into place and locks into position, the other belt, fresh from the pressure cooker and full of diamonds, swings out over the work area.

The operator raps the freshly baked sample sharply with his ball-peen hammer, freeing a black cylindrical plug about the size of a wine bottle's cork. He strikes the freed sample again and again, breaking away soft, dark grey gasketing material and revealing a hard cylindrical mass, $3/4$-inch thick and studded with tiny diamonds the size of coarse sand – perhaps thirty carats of diamonds in all. Without hesitation he pops the diamond-rich slug into a drawer with dozens more just like it. He wields a wire brush to clean off the carbide ring and inserts a new sample assembly into the belt's waiting hole. The entire process, lasting only five or six minutes, is ready to begin again.

Thirty carats per run, ten runs per hour, eight hours per day – a good worker can turn out several pounds of diamond a week. Many dozens of presses operate like that for twenty-four hours a day, three shifts a day, seven-days a week. The production of diamond is prodigious. The annual output of Ohio diamonds – thirty-three tons in 1990 alone – rivals the rest of the world's diamond production, natural and synthetic combined. In a quarter century of operations General Electric has synthesized hundreds of tons of synthetic diamonds – far exceeding the quantity of all diamonds mined since Biblical times.

But synthesis is only the first of several processes performed at the Worthington facility. Every day thousands of diamond-and-metal slugs must be processed in huge acid vats. Workers in protective rubber suits monitor the baths and retrieve piles of diamond sand from the caustic brew. Then they wash the diamonds and place them in drying ovens. The concrete floors of the chemical-preparation rooms sparkle from countless tiny diamond crystals that have worked their way into the hard surface. Worthington employees have learned to change shoes before going home at night; the acid residues and diamond grit embedded in their soles ruin uncarpeted floors.

After drying, some of the processed diamond is taken to another part of the Superabrasives factory, where the fine-grained diamond product Compax is made. There, workers mix diamond powder with metal and subject the carbon atoms to another round of high temperature and pressure. The large presses at Worthington produce thousands of diamond disks up to two inches in diameter to be used for making tools and dies.

Another area of the facility uses fine-grained diamond as the starting material for producing large, flawless single crystals. Though a relatively minor part of GE's total production, these crystals are in ever-increasing demand for electronics and research applications. Although they are exorbitantly expensive – each one monopolizes a large press for days at a time – the perfection of synthetic diamond crystals, and their ability to remove heat from sensitive electronics, is unrivaled.

The final step in preparing GE diamonds for market takes place in a small room that is perhaps the most impressive of all. All the synthetic crystals must pass through the sorting room, where diamonds are sized and weighed. It is an immaculately clean place, except for the piles and buckets of diamonds everywhere (fig. 45). One container holds 25 000 carats of perfectly sized yellow crystals. More thousands of carats stand in neat conical piles on white paper next to a black binocular microscope. Jars of black-, copper-, and honey-colored abrasives line the shelves. Only two exits lead from the sorting room: one heavily reinforced door to the factory corridors, another to a loading dock designed for armored cars.

In four decades of research, development, and marketing, General Electric's diamond making venture has gone from a few men struggling with a leaky old press to an international business grossing hundreds of millions of dollars a year. With such extraordinary success it is no surprise that GE has spawned many rivals in the diamond-making game.

Fig. 45 Harold Bovenkerk in the diamond weighing room of General Electric's Worthington, Ohio, plant, *circa* 1990. (Courtesy of Harold Bovenkerk.)

Notes

1. Letters from Armando A. Giardini to Robert M. Hazen, January 9, 1992, and January 22, 1992.
2. Details of Joe Boyd and Joe England's career and research at the Carnegie Institution were obtained through interviews with Joe Boyd. Boyd also provided copies of relevant correspondence from the 1950s.
3. Carnegie Institution of Washington, "For immediate release," January 22, 1960. (Washington, DC: 1 p).
4. Francis R. Boyd and Joseph L. England, "Minerals of the mantle." *Carnegie Institution of Washington Year Book* **59**, 47–50 (1960).
5. Herbert Strong provided his recollections of the explosion during interviews in 1991.
6. Herbert M. Strong and Robert H. Wentorf, Jr., "The growth of large diamond crystals." *Die Naturwissenschaften* **59**, 1–7 (1972). Strong and Wentorf provide an historical overview in: "Growth of large, high-quality diamond crystals at General Electric." *American Journal of Physics* **59**, 1005–1008 (1991). Many details of the history and process were provided in interviews and correspondence with Strong and Wentorf.
7. T.R. Anthony, W.F. Banholzer, J.F. Fleischer, L. Wei, P.K. Kuo, R.L. Thomas, and R.W. Pryor, "Thermal diffusivity of isotopically enriched[12] C diamond." *Physical Review*, **42**, 1104-1111 (1990).

The GE development may have been anticipated, at least conceptually, in the mid-1970s by independent researcher Russell Seitz, who was issued U.S. Patent No. 3,895,313 for concepts related to thermal conductivity of isotopically pure synthetic diamond. For background see: Eliot Marshall, "GE's cool diamonds prompt warm words." *Science* **250**, 25–26 (1990), and subsequent letters to the editor by Seitz and others: *Science* **250**, 1194–1195 (1990).
8. T.R. Anthony *et al.*, op. cit. See also R. Pool, "Ultra diamond from pure carbon-12." *Science* **249**, 28 (1990).
9. W.F. Banholzer, Physical Review B, October 1, 1991. A summary of GE's press release appears in: Ivan Amato, "GE achieves dial-an-isotope diamonds." *Science* **254**, 653 (1991).
10. A. K. Ramdas, S. Rodriguez, M. Grimsditch, T.R. Anthony, and W.F. Banholzer, "Effect of isotopic constitution of diamond on its elastic constants: [13]C diamond, the hardest known material." *Physical Review Letters* **71**, 189–192 (1993).
11. The history of Compax development was related during interviews with Robert Wentorf in 1991.
12. I thank Dr. Mark A. Sneeringer and Ms. Anne Shayeson of GE for conducting a tour of the Worthington facility on Friday, April 12, 1991. They also provided extensive promotional material on GE Superabrasive products. Additional information was provided in interviews and correspondence with Harold Bovenkerk.

CHAPTER 10

THE RIVALS

Imitation is the sincerest of flattery.

CHARLES CALEB COLTON, *Lacon*, 1820

T HE WORLD'S APPETITE FOR diamonds has become almost insatiable. Diamond mines in Africa produce nearly ten tons annually, roughly matching the output of Australia and Russia, the two other major producing areas. Slightly less than half of this production supplies the $4 billion annual gem market, while the rest provides abrasive material for industry. But this quantity pales besides the world production of synthetic diamond, estimated by GE authorities to exceed one hundred tons per year.

General Electric's success has inspired many rivals in the manufacture of synthetic diamonds; the business has grown too big and lucrative for just one player, and demand for the abrasive continues to increase each year. Every new automobile requires about one-quarter carat of diamond abrasive during its production; every truck, plane, or missile uses many times that. And diamond-tipped tools have become essential for oil drilling, stone polishing, one-hour eye glass production, and a hundred other manufacturing operations.

At the time of GE's diamond-making breakthrough De Beers had long planned to make its own synthetic diamonds in order to protect its diamond monopoly.[1] The company had established the Johannesburg-based Diamond Research Laboratory in 1947, though at the time little thought was given to the synthesis of diamond. The handful of scientists at the lab focused only on investigating the properties of natural stones. Immediately after GE's 1955 announcement, however, priorities at the De Beers facility changed, and Director of Research Dr J. F. H. Custers quickly assembled a synthesis research team.

First on board was physicist and crystal expert Henry B. Dyer, a Cambridge-educated South African who worked for De Beers, studying diamond crystals at Reading University in England. Custers also recruited D. B. Senior, a mechanical engineer from Sheffield University; Alan Blainy, who was lured away from the Harwell atomic energy research lab; and a young graduate student from Reading named C. Phaal. The project's new building, the Adamant Research Laboratory, was constructed and equipped in 1957.

The task assigned to the De Beers researchers was daunting. As Henry Dyer puts it, "Make diamond, or fail and waste three or four years of effort." None of the team members had any training in high-pressure research, and GE's public announcements gave few hints about how to proceed. Dyer recalls, "There was nothing much to go on other than the certain knowledge – not insignificant, though sometimes very depressing – that the problem could be and had been solved."

Initially, they tried a piston-and-cylinder device just to get high-pressure experience. After many trials they achieved their first taste of success in October, 1958, when a tiny diamond crystal was found in an experiment with graphite and tungsten carbide as starting materials. Out of dozens of subsequent runs, however, traces of diamond were only found three or four more times.

In late 1958 P. T. Wedepohl, a recent graduate of Witwatersrand University, joined the group and proposed an alternative device with opposed conical anvils, remarkably similar to Tracy Hall's belt. De Beers researchers claim to have developed the idea independently, though Hal Bovenkerk emphasizes that by 1957 belt specifications had been widely copied and circulated inside General Electric in the form of a blue, spiral-bound confidential document, and those plans could have leaked out easily. Construction and calibration of the new De Beers device was completed by May 1959, and the first trials were ready to begin in June.

The chemistry of diamond growth presented the other big challenge to the De Beers group. Henry Dyer, who had thought long and hard about the optimal chemical environment for growing diamond, had also concluded that nickel or iron would be an ideal solvent for crystal synthesis, and metal thus became a key ingredient in all subsequent experiments. Dyer may have had some help in reaching this conclusion. Hal Bovenkerk is quick to point out that all of GE's commercial product, which had been on the market since 1957, contained telltale traces of these critical metal solvents.

With their new press and revised chemical procedures, the De Beers team created diamonds almost immediately, though not consistently enough to begin large-scale production. "Had we patented our apparatus and process then [in the summer of 1959], the legal story would have been considerably different," Dyer laments. "But we were naive in these matters and merely went ahead with our experiments." They waited until September, when a reliably reproducible procedure was found. In a refrain repeated by all of history's diamond makers, Dyer remembers, "The euphoria when we succeeded was enormous." Without further delay the patent attorneys were brought in, but by then they were too late. GE had filed its international patents just days before.

It suddenly appeared that all the De Beers research would be for naught. There was little choice for the South African company but to go on the offensive and challenge General Electric's patents in court. No one questioned whether General Electric held valid diamond-making patents in the United States, Canada, and Great Britain; those documents had been filed and then sealed in early 1955. At issue were GE's claims to control diamond synthesis throughout the rest of the world. Could De Beers legally manufacture diamond abrasive in South Africa? If so, then by a quirk of international law they could sell their product freely to American companies, the world's largest consumers.

In the early 1960s De Beers and General Electric commenced a mammoth, six-year legal battle costing many millions of dollars – at the time the most expensive court proceedings in South African history. In the most protracted of several separate lawsuits, De Beers challenged the originality of GE's discovery and the accuracy with which they had described it. To make their case, De Beers engaged a number of expert witnesses to discredit General Electric in the eyes of South African law. Of these distinguished scientists, none was more colorful or controversial than George Kennedy.

* * *

Of the many young physicists and geologists who traveled to Harvard to learn the secrets of high pressure from Percy Bridgman and his colleagues, George Kennedy (fig. 46) was certainly the most flamboyant and outspoken.[2] He is remembered as a researcher of great creative power and *joie de vivre*.

His bold character was etched on the mind of everyone who knew him – colleagues called him outrageous, brilliant, contentious,

Fig. 46 George Kennedy, *circa* 1975, in his UCLA laboratory. (Courtesy of Art Montana.)

inspired, aggressive, and just plain rude. Some treated him with awe or fear, others with cautious respect or outright contempt. Only a few of his scientist colleagues remember him as a friend. Perhaps that isn't surprising, for Kennedy made scores of enemies with shameless public antics and unrestrained criticism of his peers.

At the end of a particularly aggressive series of long, typed letters to Joe Boyd debating aspects of pressure measurement – letters filled with insults like "stupid," "incompetent," and "dumb" – Kennedy added a conciliatory handwritten note. "Some of my lab help think I was being too heated in my arguments. It's just because I naturally argue eagerly – the real thing is that we both are trying to get at the facts. So *please* ignore any heat in my arguments. It's just my natural way – and I don't want to rewrite the letters."

Such apologies were, by all accounts, rare; even the most senior professors and science luminaries were not immune to his impolitic scorn. "Dead wood," he'd call them, not caring who overheard. And if he couldn't score a verbal victory, he was willing to settle a debate with his fists. No one seemed to enjoy these confrontations, and a good many of his contemporaries were simply scared of him.

Yet behind his outrageous behavior lay a remarkable flair for intuitive problem-solving and a keen scientific intellect, one that was all the more remarkable because of his modest upbringing on a Montana cattle ranch. Kennedy had been awarded a Harvard scholarship at the age of sixteen and became a Harvard Junior Fellow in geology soon thereafter. Following B.S., M.A., and Ph.D. degrees, and a brief stint teaching at Harvard (where he advised graduate student Joe Boyd), he joined the earth science faculty at UCLA in 1953. In decades of top-notch research he gained a paradoxical reputation as a complete wild man who was nevertheless lionized by the elite of Los Angeles society – a contentious scientific debater unhindered by professional decorum, yet a man dedicated to achieving unassailable accuracy in every measurement he made.

Everyone who knew Kennedy has amazing stories to tell. His academic research was driven by three great passions: conducting high-pressure research, collecting primitive art, and cultivating rare orchids. He once astounded colleagues by presenting three lectures at Oxford in one day – one on synthesis of high-pressure materials, one on dating pre-Columbian pottery, and one on new species of exotic orchid. Kennedy often boasted of his skill in merging the latter two passions. He successfully smuggled ancient South American pottery into the United States by covering them with piles of rare orchids in the back of his car. Customs agents focused on the flamboyant plants, which were legal imports, and overlooked the priceless art.

George Kennedy was audacious and more than a little unscrupulous. Archaeologists had long been unable to locate the source of a steady stream of valuable burial artifacts being sold by villagers from one little-studied area of Mexico. While most scientists searched for years to locate an undisturbed grave, Kennedy used a different approach. He flew by small plane to a local village, rented a jeep, and stood in the back of the vehicle waving a thousand-dollar bill and shouting in halting Spanish that the bill would go to the first person who took him to an unopened tomb. The very next morning he stood in a pristine chamber lined with precious burial artifacts. Effective as it was, his ingenious solution failed to impress the outraged Mexican authorities. One time, so the story goes, when he was stopped for his illegal exploits by Mexican police near the California border, he dutifully pulled over. Then, as the patrolman approached on foot, Kennedy floored the gas pedal in his car and escaped to the United States.

"How he ever stayed out of jail I don't know," mused Alvin Van Valkenburg, who knew him both at Harvard in the early 1950s and as a competing high-pressure researcher in subsequent years.

Kennedy's exploits in search of primitive New Guinea art would seem to rival those of Indiana Jones. On one such trip to the Maprek region he rented a dump truck and headed inland along primitive dirt roads. At each village he stopped and offered cash for ancient wooden carvings. By trip's end a few weeks later, he had amassed two trucks full of art treasures. On another expedition he rented a boat and traveled alone up the Sepic River into the most dangerous part of New Guinea's interior – a region inhabited by headhunters. Official anthropological expeditions to the area took months of planning with government assistance, but Kennedy just took off by the seat of his pants and reappeared a few days later, his boat laden with tons of priceless artifacts. He boasted in a letter to Joe Boyd, "I got most of the loot from two or three crumbling spirit houses . . . and have approximately 40 tons of stuff on its way now. Am setting up my own New Guinea village on our tennis court."

He played the game with abandon. Customs officials in New Guinea had been assured that all the carvings were less than twenty-five years old, and thus not subject to export restrictions. At the other end, in Los Angeles, equally persuasive evidence was presented that all objects were more than a hundred years old, and thus not subject to import duties. He then avoided paying income taxes by donating the least interesting objects to museums and deducting their value as charitable gifts, while his personal collection of the choicest pieces grew.

Such antics made good stories, but also angered serious archaeologists. When a group of earth scientists nominated Kennedy for membership in the prestigious Cosmos Club, a private society for intellectuals in Washington, DC, he was soundly vetoed by a lobby of incensed archaeologists.

Kennedy's extravagant lifestyle was even further embellished by his marriage to Hollywood heiress Ruth Book. Their relationship got off to a rocky start when he burned her house down while smoking in bed, but to make amends he helped acquire a magnificent home in West Los Angeles on the crest of the Santa Monica Mountains overlooking UCLA. "It was the least I could do," he explained.

He landscaped the property to include a series of water lily ponds populated by a variety of exotic water birds. As time went by he found that the ponds began to fill up with algae, and decided to enlist the services of a small South American fish that he knew thrived on the unwanted scum. Unfortunately, the fish usually died at water temperatures below about fifty degrees, and they couldn't survive winters at the Los Angeles hilltop. The ingenious Kennedy applied to the National Science Foundation for a grant to study and adaptively breed the algae-eating fish for resistance to lower temperatures. He used the NSF funds to purchase twenty-five-gallon aquariums with state-of-the-art water temperature controls, stocked with hundreds of fish. Day by day he lowered the water temperature a degree at a time until ninety percent of the fish died. The remaining ten percent, those most resistant to cold, began the next cycle of breeding. Unfortunately, his government-subsidized experiments failed to produce fish that could live in his pond.

Kennedy would try out almost any outrageous idea. He once made headlines by producing and marketing a recording of heartbeats. He claimed newborn babies, used to the sound of their mother's heart, would sleep better with the recording playing over and over again, and reportedly made a lot of money marketing it.

Joe Boyd quipped: "After associating with George Kennedy for a while it was clear you could get away with a great deal in life that I hadn't been getting away with."

* * *

Kennedy was the perfect gunslinger for De Beers' attack on General Electric. He was a leader in high-pressure research and one of the first to duplicate the diamond synthesis feat following GE's announcement. He brought his flamboyant, aggressive approach into the British courtroom as easily as he did to the staid worlds of art, orchids, and scientific debate. With blustering self-confidence, Kennedy believed that there were several possible avenues of attack to discredit the GE diamond-making process in court. General Electric had specified a range of pressures and temperatures for diamond synthesis and a variety of starting materials based on carbon plus a metal. If either of these criteria were in error, or if additional sets of viable conditions could be demonstrated, then the GE patent could be circumvented. Kennedy first focused on the question of pressure.

In the early 1950s, when General Electric scientists commenced their studies, everyone used the methods devised by Percy Bridgman to measure pressure.[3] Bridgman had devised two principal means to calibrate his high-pressure runs. In one set of experiments he made careful calculations of the pressure at which various metals were observed to undergo sudden changes in crystal structures – transformations that caused a sharp volume change and consequent jump in the piston. Bridgman reported values for bismuth, thallium, cesium, and barium, with transitions between about 25 000 and 60 000 atmospheres.

In many experiments it is impossible to use a volume change to measure pressure, so Bridgman developed a second scale based on sharp changes in the electrical resistance of those same metals. According to Bridgman, the resistance changes took place at much higher pressures than the volume changes – he thought that the volume changes and resistance changes were completely unrelated phenomena. Unfortunately, the General Electric team had to rely on resistance measurements to calibrate their pressure devices.

Today we know that Bridgman was wrong. The volume and resistance changes occur simultaneously. In the critical case of cesium metal, for example, the transformation occurs at about 44 000 atmospheres, as opposed to the 54 000 atmospheres he assumed. A pressure of 100 000 atmospheres on Bridgman's old scale is, in reality, no more than about 75 000 atmospheres on the corrected scale. For years the GE workers, who used the erroneous resistance scale, thought they were achieving pressures well above 50 000 atmospheres, which according to their calculations was more than enough to make diamonds. In fact, pressures were closer to 40 000 atmospheres, *slightly* below the graphite–diamond transition pressure. If the GE team had known about the error, they might have switched to carbide components sooner, and thus cracked the synthesis barrier years earlier. Instead, they had focused intensive effort on trying to make diamond at conditions at

which graphite was stable. They had succeeded only when the belt apparatus was fitted with carbide pistons capable of achieving higher pressure, which allowed them to push their work well into the region of
diamond stability.

George Kennedy suspected Bridgman's error and he pounced on the
chance to discredit General Electric – and Percy Bridgman in the
process. In June 1960, Kennedy, along with his graduate-student assistant Phillip N. LaMori, began to publicize a revised pressure scale.[4] He
wrote dozens of letters, proclaiming "We have very strong suspicion . . .
that [Tracy Hall's tetrahedral press] and G.E. apparatus are in gross
error at high pressures, probably as much as 30% at 100 kb [100 000
atmospheres]. . . . I would not be a bit surprised if G.E.'s 100 kb measurements actually turned out to be 70 or less."

One of the most memorable presentations of these disturbing results
occurred at the International High Pressure Conference, held at Lake
George's Sangamore Conference Center in Bolton's Landing, New
York. It was the first big international meeting on high-pressure
science, with delegations from Russia, Japan, Sweden, and other
centers of research. The conference provided the kind of forum at which
George Kennedy could be at his most outrageous.

Members of the original GE team remember with chagrin and wry
amusement the contempt that Kennedy showered on them and
Bridgman as he pointed out the "stupidity" of their measurements. He
repeatedly applied the term "honest kilobars" to his own work, implying
something less than honesty in the work of others. These attacks deeply
embarrassed General Electric and Bridgman, and offended many high-
pressure researchers who saw no need to scorn the earlier work, even if
it was in error.

Kennedy repeated these claims in the South African courtroom
where De Beers contested GE's diamond-making patent.[5] GE presented strong counter-arguments to Kennedy's ruthless attacks. Their
patents described diamond synthesis in terms of reproducible experimental procedures, not any absolute pressure scale, for exactly that
reason.

Undeterred, Kennedy also challenged GE's description of the
diamond synthesis chemistry. GE scientists claimed that the presence
of a metal like iron or nickel was essential both as a way to dissolve the
carbon atoms and as a catalyst to stimulate diamond growth. Kennedy
argued that iron served only to dissolve the carbon, nothing else. In his

capacity as consultant to De Beers, he helped gather more than twenty affidavits from scientists around the world who agreed with his point of view. To counter that challenge, Bob Wentorf performed a series of experiments showing that many molten substances, such as silver chloride and cadmium oxide, dissolve carbon at diamond-making pressures, but they don't make diamonds.

Such subtle scientific arguments made for a complex and protracted legal battle, and George Kennedy was only the most colorful of more than a dozen expert witnesses De Beers put on the stand. Thus the trial dragged on for many weeks until, finally, presiding Justice Roberts had heard all arguments. He was expected to deliver his judgement within a few months, but he died suddenly before his decision was ever written. After a delay of another year, the entire trial had to be reheard. The second justice ruled in GE's favor, but De Beers immediately appealed, and so the case continued.

* * *

While De Beers attacked the validity of GE's process patents, a second round of lawsuits and countersuits focused on the belt apparatus. De Beers was so determined to join the synthetic diamond game that it had secretly constructed its own belt-type machines, which some scientists suspected were based on GE patent details. Belt-type devices had been put to work at the well-guarded De Beers plant in Springs, South Africa, just a dozen miles south of the Premier Mine. De Beers successfully operated the plant in secrecy for several years until they foolishly published a photograph of their operation in promotional literature. The photograph showed the distinctive tapered anvils of a belt-type device.

With unequivocal evidence of De Beers' patent infringement, GE sued the South African giant. In a standard legal ploy, De Beers countersued, claiming that GE had themselves revealed details of the belt prior to filing their patent, thus invalidating their claim for protection. Again the lines were drawn and the legal battle began.

In late 1960 the high-pressure experts, Tracy Hall, Francis Bundy, Hal Bovenkerk, and Armando Giardini, who had left the Army for a professorship at the University of Georgia, were flown to South Africa for the main event.[5] General Electric lawyers had guessed that the De Beers strategy would be based on a key point of patent law: if General Electric had released details of the belt apparatus before the patent date, then their claim was invalid.

Hal Bovenkerk, the first GE witness scheduled to testify, suspected that the De Beers defense would highlight a 1954 GE publicity photograph of the thousand-ton press with a bit of a circular girdle showing. By publishing the photo, De Beers would argue, General Electric had forfeited its rights. What no one outside of the GE team knew was that at the time of the photograph it was Strong's cone apparatus, not the belt, that was mounted in the big press. Bovenkerk played the cat-and-mouse game beautifully, answering four days of intense questions with the traditional "Yes, my Lord" and "No, my Lord." When the De Beers lawyer at last, with a flourish, introduced the key photograph, Bovenkerk was ready.

"Is this a photograph of the General Electric press?" the lawyer asked.

"Yes, my Lord." Bovenkerk paused, and then, anticipating the next question, added "But it's not a belt!" The De Beers case was in ruins.

General Electric seized the advantage and pressed De Beers to settle the case. Within a few days of Hal Bovenkerk's testimony, De Beers purchased limited rights to make diamond for a sum that, though not disclosed, was rumored to be about $25 million, many times the total General Electric investment in diamond research. In keeping with industry practice, De Beers also agreed to pay royalties on its diamond production.

Armando Giardini, who never had to testify, remembers the South African trip as a great month of sightseeing, though he spent the month of his return recovering from a nasty bout of tapeworm. Herb Strong and Bob Wentorf, who were scheduled to testify at the second trial, remember the terse, happy telegram: "CASE SETTLED. DON'T COME." Herb Strong and his wife traded in their tickets and took a vacation to Europe instead.

Who won the legal battle? General Electric partisans saw the result as a satisfying and long overdue victory for the American corporation. Members of the De Beers group claim that they gained the upper hand, and that General Electric's offer of a licensing agreement was merely a ploy to avoid eventual legal defeat – a contention that gives GE partisans a hearty laugh. In one sense it makes little difference, for both companies have since made a fortune selling synthetic diamonds.

* * *

Although General Electric and De Beers have remained the major players in the diamond synthesis drama, many others have tried to get into the act. The irrepressible George Kennedy, who approached everything he did with passion, became obsessed with making diamonds. He was absolutely convinced that the most versatile high-pressure device was the simple piston-cylinder, and that conviction drove him for almost two decades.[6]

Belt and tetrahedral anvil devices have limited sample volumes; eventually, as pressure is raised, the opposed anvils must come into contact. But a piston-cylinder, Kennedy argued, has the potential for unlimited stroke and therefore huge sample volumes. He also believed that the simpler design of a perfect piston in a perfect hole allowed the best control of temperature and pressure, as well as the offering the easiest machining and replacement of parts.

George Kennedy was too late to invent diamond making, but he reasoned that if he could make twice as many diamonds with half GE's effort, he would become rich beyond his dreams. He was also smart enough to use somebody else's money in the expensive effort. For almost twenty years Kennedy flitted from corporation to corporation in his quest for diamonds. The first abortive attempts were sponsored by Hughes Corporation, the great aircraft builders in Los Angeles. Kennedy oversaw the construction of a press and a piston-cylinder device fashioned from high-grade Maraging steel, with an elegant water-cooling system. The machine could reach diamond-making conditions – just barely – but what Kennedy didn't realize was that Maraging steel corrodes quickly in hot water. Hughes canceled the project when the press exploded, nearly killing two workmen.

Undeterred, Kennedy quickly arranged a project with Teledyne in the mid-1960s and built another piston-cylinder rig, this time with the aid of consultants Ivan Getting and John Heygarth, both of whom worked in Kennedy's high-pressure laboratory at UCLA. Getting had employed carefully aligned carbide components, and thus succeeded in sustaining diamond-making conditions at the UCLA facility. Teledyne decided to take advantage of that expertise. They hired the UCLA workers and authorized a duplicate copy of the university's machine. Getting's magic touch paid off, for they readily synthesized diamonds. But in the process, the pros and cons of the piston-cylinder became more apparent. On the positive side, Kennedy's Teledyne group could obtain large diamond yields with a graphite and metal sample four inches long and a half-inch in diameter; each successful run produced more than one hundred carats. But the costly carbide pistons broke constantly, often after only one or two runs, and the group was soon carting away shattered pieces of carbide in wheelbarrows. By the mid-1970s, after almost a decade of expensive research and no clear prospect of a commercially viable process, Teledyne threw in the towel.

Without hesitation Kennedy scouted out his next backer. The choice came down to Norton, with their expertise in abrasives, or Kennametal, a company that had extensive experience with carbide products. Getting and Kennedy settled on Kennametal, and a new project was underway within weeks. Getting was given the design chore: construct a piston-cylinder device with a one-inch diameter bore and a six-inch-long sample chamber that would fit into the thousand-ton press already at Kennametal's Latrobe, Pennsylvania, plant. Those specifications provided room for a sample five inches long and three-quarters of an inch in diameter – enough graphite to make more than three hundred carats of diamond at a shot.

Though progress was slow, Kennedy maintained management's enthusiasm with his unbridled optimism. He mesmerized Kennametal executives with promises of piles of diamonds just around the corner. "It's the best idea since the invention of sex!" he would shout. He did have some cause for optimism: Getting's press did work wonderfully well. The principal drawback was the team's lack of expertise in the chemistry of crystal growing. They could make diamond, sure enough, but never with the consistent size and properties essential for commercial use. Perhaps sensing a cooling of corporate enthusiasm, Kennedy jumped ship in 1978, and Kennametal gave up the project within a year of his defection.

For his fourth and final foray into commercial diamond making, Kennedy turned to a local firm, Research and Design Associates (RDA), a Los Angeles think tank, which focused primarily on consulting for the military. Although RDA had no production engineers, no marketing management, and no sales force, Kennedy was a friend of the top RDA management people, and he made them a truly mind-boggling proposal. His grandiose idea was to construct a two-inch-diameter piston-cylinder in a three-thousand-ton press, large enough to churn out more than a thousand carats – almost a quarter of a pound – at a time. Kennedy's vision was not limited to diamond grit; he envisioned mammoth single crystals, diamond semiconductors, and special diamond components for use in fancy lasers.

But George Kennedy never lived to see the dream fulfilled. At the time of his death in 1980, RDA had authorized only a scaled-back prototype with a half-inch diameter cylinder – a device that sat rusting for more than a decade in a Los Angeles warehouse.

Throughout this period of high-powered corporate consulting in the 1960s and 1970s, George Kennedy engaged in his own frenetic diamond work at UCLA. Though trained in earth science and dedicated to understanding the origins of rocks, his lab research became more and more focused on diamonds. He supervised hundreds of runs, each with a slightly different recipe or synthesis conditions. Some say he appeared to be guided more by hunches than by any systematic plan, and several coworkers sensed an air of desperation in his efforts. In almost twenty years of research, Kennedy produced diamonds in hundreds of experiments but did not see one carat of his synthetic diamond sold commercially.

* * *

Several foreign ventures have fared better. Ireland became a major diamond-producing country in 1962, when De Beers shipped twenty-five of its original presses to a new plant, the Ultra High Pressure Unit (Ireland) in Shannon, which now synthesizes fifteen tons of abrasives annually. General Electric established a diamond factory in Dublin, Ireland, in the early 1980s. About half the size of the Worthington facility, it too produces about fifteen tons of crystal per year. A De Beer's diamond-making plant on the Isle of Mann, rumored to be exclusively for single-crystal synthesis, is said to create some of the most perfect diamonds ever seen, including a 14.2–carat monster – perhaps the largest ever grown. Diamond production in the rest of Europe, concentrated in France, Germany, and Sweden (an outgrowth of von Platen's original ASEA effort), approaches twenty tons annually.

The former Soviet Union committed vast resources to diamond synthesis, though only recently have details been forthcoming about their success.[7] Statistics on diamond production were decreed a state secret by the Soviet government in 1956, though they are believed to have operated major diamond production or research facilities at Poltava and Lvov in the Ukraine, Erevan in Armenia, and St. Petersburg in Russia.

Today, several Asian countries, with their growing industrial capabilities, have become active in developing diamond-making factories. Most of the Japanese synthetic diamond production is in the form of abrasive manufactured by Tomei, but remarkable gemstones of more than a dozen carats have been produced by Sumitomo. The Chinese, by contrast, have established their own diamond-making infrastructure, a remarkable cottage-industry approach unique in the world. There are reputed to be dozens of individual diamond-making presses scattered around the country. Tracy Hall recalled a 1988 tour of one extraordinary site, a mom-and-pop operation featuring a main building for jade carving, and a backyard shed with a thousand-ton press.[8] The juxtaposition of the ancient carving art with a modern press seemed strange to Hall, but the system has its own logic. Jade carving requires diamond-tipped tools (one word for diamond in Chinese may be translated as "cutting stone"); whenever a new abrasive supply is required, the carvers go out back and whip up another batch. Not all Chinese production occurs on such a small scale, however. They are said to have purchased one European five-thousand-ton press, to serve as a model to duplicate the device for a large-scale diamond-making factory.

Of all the world's diamond-making efforts, none reveals more about the gem's temptations than the schemes of Chen-Min Sung to establish the industry in Korea. During the 1980s Sung played a unique and vividly sinister role in the diamond synthesis story.[9] He learned high-pressure technology at MIT in the mid-1970s and eagerly accepted a position as Engineer at GE's Worthington plant in the late 1970s. Sung began as an enthusiastic and ambitious researcher who wanted to improve diamond-making procedures, but, he claims, the management was unresponsive to his ideas. Eventually, disillusioned but rich with GE secrets, Sung accepted an attractive offer from Norton's Vice President in charge of research, Peter Bell. At Norton, Sung was given much more flexibility in research and development, and free access to their proprietary technology. It now appears that he abused that trust. Adding his knowledge of Norton's manufacturing and processing methods to his inside information on GE superhard materials, Sung is believed to have secretly sold confidential technology – press designs, synthesis processes, and other confidential data – to a Korean firm.

In 1987 Chen-Min Sung was arrested by the FBI and soon thereafter convicted of transporting stolen property across state lines. His prison sentence now over, Sung and his Korean company face civil law suits from General Electric and Norton. Chen-Min Sung's actions shocked and dismayed the high-pressure community, which for the most part adheres to a strict ethical code. But, as Hal Bovenkerk has observed, "Scientists can be crooks as well as anybody else."

* * *

"So many people have entered into diamond making, only to lose money," Tracy Hall laments. But Hall can tell a very different story. Megadiamond, a company he helped found, survives to this day. Though it only claims a few percent of the market, Megadiamond remains GE's biggest domestic rival.

From the moment of Hall's defection in mid-1955, he became a subject of concern to General Electric. No one outside the company knew more secrets, or had better cause to share those secrets with others. De Beers, in particular, posed a tremendous threat; with Hall's knowledge they could have easily duplicated all General Electric's apparatus and processes. Yet while Tracy Hall was not about to sell trade secrets, he did have business in his blood. After independently inventing the tetrahedral press, he almost immediately began to build and sell high-pressure devices. In his Provo, Utah, workshop Hall fashioned a dozen big tetrahedral presses of his own design and sold them to government and university labs. But he soon realized that the four-anvil design was not the best for making diamonds. It was possible to get even more diamond-making volume by using six anvils, each pressing on a face of a cube.[10]

Cubic-anvil presses have become the rage in modern high-pressure research. They represent a kind of marriage between von Platen's split sphere, which was awkward to use, and Hall's tetrahedral anvil, which could be tricky to align. The cubic-anvil device features six carbide anvils, arranged in three opposed pairs at right angles to each other. The resulting sample chamber is a cube, just like von Platen's, with each of the six cube faces defined by one of the six anvils (fig. 47). In the early 1960s the cubic-anvil press was an idea waiting to happen, and several groups hit upon the design more or less at the same time. Tracy Hall's innovation was a complex arrangement of guide pins that kept all six anvils precisely aligned while pressure was raised or lowered.

While General Electric continues to use the belt to make diamonds, several small American operations rely on cubic-anvil presses of Tracy Hall's design. U.S. Synthetic Corporation (USS) in Provo's East Park industrial center is a prime example of a small-scale producer.[11] Its

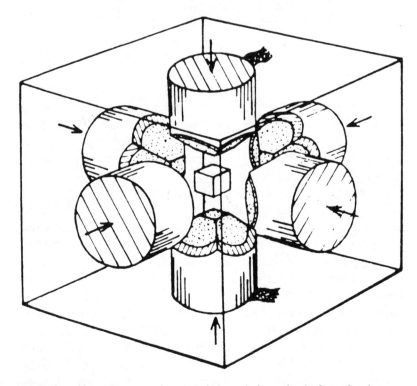

Fig. 47 The cubic-anvil press employs six carbide anvils directed at the faces of a cube-shaped sample. This design, which accepts a relatively large sample volume, is preferred by many diamond-making companies and high-pressure researchers. (Courtesy of A. A. Giardini.)

facility could hardly be more different from General Electric's vast Worthington plant. At USS a dozen workers occupy one section of a module in an unimposing village of one-story corrugated metal buildings. On the outside, the simple shelters are nondescript except for occasional company logos on the doors – a modest "USS" in an oval frame next to a plain door is all that marks the corporation's main entrance. The door opens onto a small reception area with tough thin carpet and spartan furnishings. A narrow corridor leads past cubical offices to the production area.

The employees of USS have been making sintered diamond components for rock drills since 1978, when the company split off from Megadiamond. Diamond bits for rock tunneling, excavating, and deep drilling constitute one of the major markets for synthetic diamond makers. The starting material is uniform, fine-grained diamonds no more than a thousandth of an inch across. This diamond powder, purchased in bulk from GE, is pressed into bullet-shaped pellets, which are subjected to diamond-forming conditions at the USS plant.

This is no-frills diamond making. Clean concrete floors, unadorned metal walls, and minimal fluorescent lighting create a no-nonsense environment well-suited for the work. A dozen employees feed and maintain two cubic-anvil presses, designed and built to Tracy Hall's specifications. The USS presses, crafted from gleaming stainless steel, stand eight-feet tall, alien in their massive symmetry. They operate twenty-four hours a day, six days a week, and churn out 180 runs each on a good day (fig. 48).

On a nearby set of metal shelves one-inch cubes of rusty red wonderstone are stacked like a pile of children's blocks. The press operator selects a cube that has been drilled through the center and packed with a sample of diamond powder. He places the assembly at the center of the press's six massive carbide anvils, each aimed at one of the sample's six flat faces. With inexorable force, the cube is crushed in the cubic vise, and then the heat is applied. Electric current passes through two of the opposed anvils and through the sample itself, heating the diamond powder to 1500°C while it is subjected to 60 000 atmospheres. The operator retreats behind a protective shield to begin the pressurization; once every few hundred runs – perhaps once or twice a week – an anvil will break, and at such times it's best to be behind something solid.

It only takes a few minutes for temperature and pressure to do their work. The wonderstone cubes become noticeably smaller and turn grey from the transforming heat. The press operator attacks the assembly with a hammer, easily shattering the grey shell and revealing a dense diamond core – a bullet-shaped drill piece, ready to attach to a drill bit that will slice through thousands of feet of solid rock.

From the modest beginnings in 1955, when the world's supply of synthetic diamond could have been dispersed by a sneeze, to today's thriving industry, the diamond makers have made dramatic strides. Approximately one hundred tons of diamond are synthesized every year, providing almost nine out of every ten carats used in the world. In four decades synthetic diamond production has exceeded the historical output of all the world's mines several times over.

Yet even that extraordinary output may eventually pale beside the astonishing advances of the next generation of diamond makers.

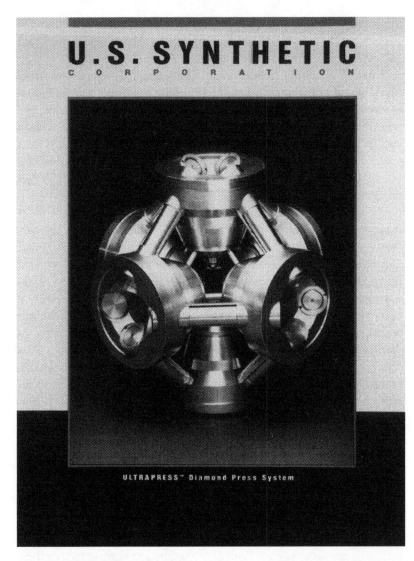

Fig. 48 The cubic-anvil Ultrapress, designed by Tracy Hall, features six pistons that compress a cube-shaped sample. U.S. Synthetic Corporation in Provo, Utah, markets the eight-foot-tall press and manufactures diamond inserts for rock-drilling bits. (Courtesy of U.S. Synthetic Corporation.)

Notes

1. The history of De Beers early diamond synthesis efforts was provided in letters from Dr Henry B. Dyer, from January and February, 1992. Dyer was then Managing Director of De Beers Industrial Diamond Division. Additional information was supplied in letters and interviews from Francis R. Boyd, Hal Bovenkerk, Henry O. A. Meyer, Herbert Strong, and Robert Wentorf.

2. Many of his former colleagues have shared reminiscences of George Kennedy. I thank Peter Bell, Francis R. Boyd, Gary Ernst, Ivan Getting, Julian Goldsmith, Alvin Van Valkenburg, and Hatten S. Yoder, Jr. for their interviews and letters. Boyd, in particular, provided revealing files of letters from Kennedy spanning the years 1953 to 1979. Details of Kennedy's life are provided by: Art Boettcher and Robert C. Newton, "Dedication to special issue honoring George Kennedy." *Journal of Geophysical Research* **85**, 6901 (1980).

3. Percy W. Bridgman, *The Physics of High Pressure*. (London: G. Bell and Sons, 445 pp., 1949 edition). See, especially, Chapter 3, "The Measurement of High Pressure." Details of the transition pressures for Bridgman's fixed points are scattered throughout his publications. See, for example, P. W. Bridgman, "The effect of pressure on the bismuth tin system." *Bulletin of the Society of Chemistry, Belgium* **62**, 26–33 (1953).

4. G. C. Kennedy and P. N. LaMori, "Fixed pressure points for calibration of high pressure." *Geological Society of America Bulletin* **71**, Part 2, 1903 (1960).

5. Details of the South African legal battles between General Electric and De Beers were provided in letters and interviews from Hal Bovenkerk, Henry Dyer, Armando Giardini, Tracy Hall, Herbert Strong, and Robert Wentorf. In addition, Francis R. Boyd provided copies of South African newspaper reports of the trials.

6. George Kennedy's effort to develop a commercial diamond synthesis operation were chronicled in several interviews with his long-time associate Ivan Getting.

7. Aspects of Russian diamond synthesis efforts are chronicled in: R.C. Devries, A. Badzian, and R. Roy, "Diamond synthesis: The Russian connection." *Materials Research Society Bulletin* **21**, 65–75 (1996). A complementary discussion of post-Soviet diamond mining and distribution policies is presented in: Andrew R. Bond, Richard M. Levine, and Gordon T. Austin, "Russian diamond industry in state of flux." *Post-Soviet Geography* **33**, 635–644 (1992)

8. Details on Chinese diamond synthesis were provided during interviews with Tracy Hall.

9. Peter Bell and Hal Bovenkerk provided interviews and documentation on the case of Chen-Min Sung.

10. Tracy Hall obtained Patent No. 3,159,876, "High-pressure press," on December 8, 1964, for several new multi-anvil press designs, including the cubic-anvil press.

11. I thank Tracy Hall and William Pope for interviews regarding commercial diamond synthesis and for granting access to the USS Synthetic plant in Provo, Utah.

CHAPTER II

THE NEW DIAMOND MAKERS: DIAMONDS BY EXPLOSION

> As the last remaining dinosaurs neared their end, a dusting of micro-
> scopic diamonds apparently fell from the sky, according to a scenario pro-
> posed by two Canadian researchers who found extremely tiny diamonds
> in 65-million-year-old rocks from Alberta.... Over the last decade, scien-
> tists have accumulated evidence that a meteorite or comet struck Earth at
> that time and caused the extinctions. The researchers suggest the
> diamond dust was either brought to Earth by an impacting meteorite or
> created during the high-pressure collision.[1]
>
> *Science News*, September 5, 1991

IN THE ROUGHLY ROUNDED mountains of Pennsylvania's
Appalachian Mountains geologists have discovered rich mines of one
of the most sought after of all geological treasures. Deep into the earth
men have carved their caverns – drilling, blasting, and moving countless tons of rock. These prospectors seek neither precious metals nor
gemstones; they labor for gravel. Gold and silver may be good for
making trinkets, but with crushed stone you can build a nation.

The Madison Formation of western Maryland and central
Pennsylvania makes great gravel. Its massive blue-grey limestone beds,
remnants of an ancient coral reef, blast free into neat layers a foot or two
thick, a perfect size for transport to the steel-jawed rock crusher. Sieve-
like sorters guide well-sized rock to conveyor belts, each leading to a
great conical gravel pile: fist sized for drainage, walnut sized for septic
systems, pea sized for driveways and walkways.

The mining strategy is simple, but the results are epic. First, you
must find a mountain where limestone layers appear near the base. The
operation begins by cutting away a great pit against the mountain's
flank, exposing a sheer wall of limestone. Operators situate the crushing and sorting equipment close to this pit for easy access. Miners
tunnel deep into the mountain, removing every bit of limestone except

for a few widely spaced square rock pillars, four feet on a side and several stories tall, that support the vast underground chambers. Imposing fifty-foot-high ceilings and pavement-smooth rock floors define the top and bottom of what was once a solid limestone layer. Mammoth trucks with wheels taller than a man ferry freshly broken rock from the immense artificial caverns to the outside world. Deeper and deeper the operation extends, sometimes the better part of a mile laterally beneath the wooded slopes.

Eventually a limestone mine must give out; you can only undercut a mountain so far before the risk of collapse becomes too great. The deeper you excavate, the more costly the operation. But abandoned limestone mines are not without their uses. The open pits are meccas for an army of amateur fossil hunters, who comb the weathered rock rubble for beautifully preserved shells of strange, extinct sea animals. The ancient reef rock breaks with a pungent sulfur smell – you can breathe in the atoms of life entombed a third of a billion years ago.

One spent Madison quarry, about 50 miles southeast of Pittsburgh, Pennsylvania, serves a very different purpose. This mine, with its carefully guarded gates and prominent "No Trespassing" signs, serves as a chamber for making diamonds. In this underground limestone mine Pennsylvania workers pack pipes full of graphite and metal powders, and then surround the pipes with tons of explosives. When detonated, each massive explosion creates the tremendous temperatures and pressures needed to convert much of the graphite to a solid mass of submicroscopic diamond crystals.

* * *

Almost all the world's natural diamonds grow deep beneath us in the hot, dense rocks of the earth's mantle. There, carbon atoms coalesce over eons to form magnificent gems. But nature has found another way to play the diamond game. Every so often – perhaps once every few thousand years – the earth's gravitational field captures a massive chunk of rock from outer space. The extraterrestrial boulder picks up speed as it approaches the planet and, plummeting through the atmosphere at several miles per second, the rock surface glows white hot and vaporizes in a spectacular trail of fire. At last the careening mass smashes into the planet's surface in a terrible explosion, more powerful than a fleet of nuclear missiles.

Most large meteorites hit deep water (oceans cover more than two-thirds of the earth's surface), but perhaps one in three or four strikes land or shallow water, where the projectile blasts and craters rock and soil in an instant of unimaginable pressure and temperature. In such a catastrophe rocks are shattered into an exotic dust of high-pressure minerals. Common quartz changes into coesite or even denser forms, and it is even possible, under just the right circumstances, that everyday carbon compounds might convert to diamond. Everywhere there is life there is carbon, ready to be transformed instantaneously by such an impact into microscopic diamond crystals.

Layers of natural diamond dust may reveal the scope of epic catastrophes in the earth's past. Recently, Canadian geologists David B. Carlisle and Dennis R. Braman extracted microscopic grains of diamond in a layer of sediments 65 million years old – rocks deposited at the very end of the age of dinosaurs.[2] The diamond dust is so fine-grained that many scientists suspect they were caused by a meteorite impact. Did those diamonds form in the same collision that caused the extinction of dinosaurs and countless other life forms? While the verdict is still out, a majority of scientists suspect that such an impact propelled huge quantities of fine dust into the atmosphere, blackening the skies for months and profoundly disturbing the growing cycles of plants, as well as the animals that depended on them for food. As the diamond-bearing dust settled, it left a telltale layer of fine sediment in 65-million-year-old rocks around the world.

Scientists discovered meteoritic diamonds thirty years ago at the famous Canyon Diablo site at Meteor Crater, Arizona – the same locality that first yielded coesite, one of the high-pressure forms of beach sand. Unlike South African or Brazilian stones, these crystals are generally small, typically no larger than fine sand grains – a hundredth of an inch across. While most of the the Canyon Diablo particles display all of the characteristics of ordinary, fine-grained diamond, meticulous analysis of the material late in 1966 contributed to an extraordinary find: "hexagonal diamond," a new crystal form of carbon different from graphite or diamond. Apparently almost thirty percent of the meteorite diamond, including crystals up to a few hundredths of an inch across, occurred in this intriguing form.

The find was newsworthy, but not entirely unexpected. As early as 1962 scientists had predicted the existence of hexagonal diamond, and in the mid-1960s General Electric scientists had detected traces of the new material (they called it "delta-carbon") in some synthetic runs at pressures above 130000 atmospheres.[3] In fact, Francis Bundy and coworkers were able to make nearly pure hexagonal material by subjecting carefully oriented graphite crystals to such high pressures.

Upon careful examination, the GE team realized that the natural meteoritic material, their synthetic delta-carbon, and the predicted hexagonal diamond were identical. Mineralogists, ever searching for new species, pounced on the find and gave it the name "lonsdaleite," in honor of British mineralogist Dame Kathleen Lonsdale, who had contributed much to the study of natural diamond crystals.[4] In many respects lonsdaleite and diamond are remarkably similar. They have virtually the same high density and almost the same hardness. But their diagnostic x-ray patterns, while similar in some ways, are nevertheless quite distinct, indicating significant differences in their atomic arrangement.

The relationship between the normal cubic diamond structure and that of the new hexagonal form, both of which are built from pyramids of four carbon atoms, is hard to describe but easy to demonstrate with a bunch of identical spheres (try this using marbles or oranges). It turns out there are two simple, systematic ways to stack up layers of spheres. The first layer is easy – just pack the balls close together so that each one touches six others (fig. 49). This arrangement is called "close-packing" and it's often used for efficient shipping of all kinds of round objects from soup cans to sewer pipe.

The second layer is also straightforward. Build it by placing one ball in a depression formed by any three adjacent balls in the first layer. You'll notice that there are lots of spots to choose from, but that it doesn't matter at all which one you select. From then on the second layer is formed just like the first, with close-packing of balls.

Next comes the tricky part. Where do you start the third layer? You must begin this layer by selecting a depression formed by three balls in the second layer, but now there are two different kinds of depressions to choose from. Look carefully and you'll see that half of the sites are directly above the center of a ball in the first level, while half are centered over two balls in that layer. If you choose to place a ball in a depression that is directly above one in the first layer you end up with an alternating arrangement, often described as "ABABAB . . ." in the literature of

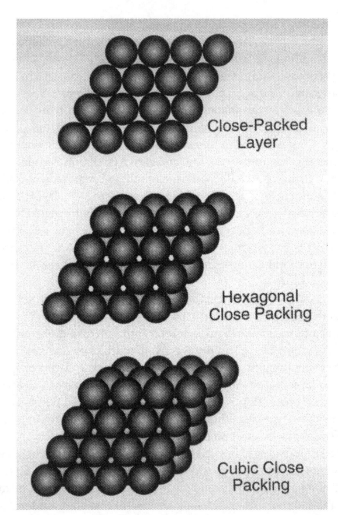

Fig. 49 In a close-packed layer, each ball touches six others. A second layer fits only one way on the first, but there are two different ways to place the third layer: either directly over the balls in the first layer (hexagonal close packing) or over the spaces between balls in the first layer (cubic close packing).

crystals – the result is hexagonal close packing. A quite different pattern arises if you make the other choice – it creates an "ABCABC . . ." sequence of layers known as cubic close packing.[5]

Diamond and lonsdaleite differ simply in the way layers of carbon atom pyramids are stacked. Diamond adopts ABCABC cubic close packing, while lonsdaleite displays ABABAB hexagonal close packing. Under most natural circumstances the cubic arrangement prevails, but under special conditions, such as the almost instantaneous shock of a meteorite impact, lonsdaleite also occurs.

While most researchers agree that the Canyon Diablo diamonds formed out of carbon atoms contained in the meteorite itself, two rival theories compete to explain their origins.[6] One group contends that Canyon Diablo diamond represents transformed meteoritic graphite: they say that black graphite grains originally embedded in the iron-rich body were converted to diamond by the pressure and temperature of impact. Others disagree, and suspect that most meteoritic diamond was formed in the vacuum of space, where they believe carbon atoms condensed directly into its denser form. The debate continues, but scientists have found that both theories – explosive shock and vapor deposition – can be used to make diamond.

* * *

Humans are quite adept at making explosions. The highest pressures available on earth come courtesy of the military-industrial complex, whose high-velocity projectiles and nuclear weapons can produce many millions of atmospheres pressure, even if only for a fraction of a second. Paul S. De Carli, scientist at the Stanford Research Institute in Menlo Park, California, and his visiting colleague John C. Jamieson, geology professor from the University of Chicago, first attempted to duplicate nature's violent impact feat in mid-1959.[7] They packed explosives behind a solid plate of aluminum metal and detonated the device with an electrical pulse. With a deafening boom, the explosively propelled metal plate plowed into a sample of graphite at supersonic speed, generating pressures greater than 300 000 atmospheres – albeit for only a millionth of a second. In experiment after experiment the shock wave transformed graphite into fine-grained diamond.

To some industry researchers, the success of De Carli and Jamieson suggested an alternative route to commercial diamond synthesis. Du Pont was America's leader in explosives technology, and they applied their expertise to a number of commercial efforts, including explosive hardening of steel, explosive cladding of metals, and explosively driven rivets. In the mid 1960s they decided to try their hand at explosive diamond making, and for more than two decades they have produced millions of carats of diamonds annually in a process reminiscent of the early efforts of Hannay and Noble.

Although it might seem that any big explosion can produce diamonds, the process is much more subtle and complex.[8] As a shock wave passes through graphite, the intense heat and pressure can indeed create diamond, but since the pressure drops rapidly as the shock wave passes the crystals can just as easily convert back into graphite while still hot. The key is to dissipate the heat as quickly as possible so that the high-pressure material doesn't have a chance to change. Du Pont originally employed a special starting mix of iron carbide with interspersed graphite particles for this purpose, though they eventually switched to a graphite–copper mixture. In each case, the metal helps to diffuse the heat. Originally, the Du Pont explosive process produced the hexagonal diamond, lonsdaleite, along with cubic diamond, but more recent modified synthetic runs only produce the preferable cubic diamond.

Technicians work deep underground in a cathedral-like limestone chamber, illuminated by giant flood lights that cast strange shadows into the dark recesses of the mine. Every small sound echoes eirily in the still air, which is always cool and dry. In a typical run the workmen pack a thin-walled cylindrical product tube with graphite and metal powder. The ends of that six-foot-long inner pipe are plugged shut, and the assembly is loaded into a thick-walled steel driver tube with an inner diameter somewhat larger than the product tube. This assembly, in turn, is placed in a sturdy culvert that has been filled with tons of Du Pont explosives. Once the loading operation is completed, all personnel are transported to safety and the explosives are detonated with a blast that rocks the limestone mountain.

The explosion collapses the driver tube around the sample, generating pressures on the order of 200 000 atmospheres – enough to convert the graphite to extremely fine-grained diamonds. The blasted and deformed sample assembly is transported by truck from the mine site in southwestern Pennsylvania to the Du Pont processing plant in Gibbstown, New Jersey. There the contorted diamond–metal mass is dissolved in acid, and the microscopic abrasive powder is cleaned, graded, packaged, and shipped.

The most striking feature of Du Pont's synthetic diamond, trade named "Mypolex," is its extremely fine grain size – typically less than a millionth of an inch – which makes it ideal for polishing gemstones, including extremely hard stones such as sapphires, rubies, cubic zirconia, and gem diamonds themselves.9 Other than Mypolex, the finest-grained natural and synthetic diamond particles commercially available are about one ten-thousandth of an inch in diameter. That's extremely small, but still large enough to leave tiny scratches and grooves – imperfections that diminish the mirror-flat surface of a finely polished gem.

Du Pont synthesizes only about two million carats (less than half a ton) of Mypolex per year, and the product costs significantly more per carat than other synthetic diamond abrasives. Nevertheless, Mypolex enjoys a steady market because it is uniquely suited to high-precision polishing applications.

And so, a hundred years after Hannay exploded bomb after bomb to no avail, humans have learned to harness the power of explosives to make diamond.

Notes

1. "Diamonds not a dinosaur's best friend." *Science News* **140**, 156 (September 5, 1991).
2. David Brez Carlisle and Dennis R. Braman, "Nanometre-size diamonds in the Cretaceous/Tertiary bounadary clay of Alberta." *Nature* **352**, 708–709 (22 August 1991). See also: E. G. Nisbet, D.P.Mattey, and D.Lowry, "Can diamonds be dead bacteria?" *Nature* **367**, 694 (1994).
3. R.E. Hanneman, H.M. Strong, and F.P. Bundy, "Hexagonal diamonds in meteorites: implications." *Science* **255**, 995–997 (1967). See also: F.P. Bundy and J.S. Kasper, "Hexagonal diamond – a new form of carbon." *Journal of Chemical Physics* **46**, 3437–3446 (1967); Kathleen Lonsdale, "Formation of lonsdaleite from single-crystal graphite." *American Mineralogist* **56**, 333–336 (1971).
4. Lonsdaleite was first described in: Clifford Frondel and Ursula B. Marvin, "Lonsdaleite, a hexagonal polymorph of diamond." *Nature* **214**, 587–589 (1967).
5. For a more detailed discussion of close packing in crystals see: Joseph V. Smith, *Geometrical and Structural Crystallography*. (New York: Wiley, 450 pp. 1982).
6. Hypotheses regarding origins of diamonds in meteorites are reviewed by: S.S. Russell, C.T. Pillinger, J.W. Arden, M.R. Lee, and U. Ott, "A new type of meteoritic diamond in the enstatite chondrite Abee." *Science* **256**, 206–209 (10 April 1992); S.S. Russell, J.W. Arden, C.T. Pillinger, "Evidence for multiple sources of diamond from primitive chondrites." *Science* **254**, 1188–1191 (22 November 1991); Michael E. Lipschutz and Edward Anders, "The record in meteorites – IV Origin of diamonds in iron meteorites." *Geochimica et Cosmochimica Acta* **24**, 83–105 (1961).
7. Paul S. DeCarli and John C. Jamieson, "Formation of diamond by explosive shock." *Science* **133**, 1821–1823 (9 June 1961).
8. Niles F. Bailey, "Explosion synthesized polycrystalline diamond powder, its manufacture, characteristics and use." In *Proceedings: Diamonds in the 80s, a Technical Symposium, 13–15 October 1980*. Chicago, Illinois: Industrial Diamond Association of

America, 8 pp. (1980). See also: Jonathan Beard, "Explosive mixtures." *New Scientist*, 5 November 1988, 43–47.

9. O.R. Bergmann, N.F. Bailey, and H.B. Coverly, "Polishing performance of poly-crystalline diamond produced by explosive shock synthesis." *Metallography* **15**, 121–139 (1982). See also: Niles F. Bailey, "Unique values of Du Pont MYPOLEX diamond powder in polishing superhard materials." Wilmington, Delaware: E.I. du Pont de Nenours & Company, 5 pp. (1988).

CHAPTER 12

THE NEW DIAMOND MAKERS:
DIAMONDS FROM A VAPOR

Diamonds are going to be everywhere. They'll be in pots and pans, on drill
bits and razor blades, in copying machines and hard discs.[1]
JOHN C. ANGUS, 1990

TODAY, HIGH PRESSURE IS THE key to most commercial diamond
synthesis, but that situation may soon change. A new generation of
diamond makers have learned that under the right conditions flawless
gemstone layers will form from hot, vaporized carbon atoms at
low pressure. Development of large-scale commercial processes has
remained elusive. But, if the dreams of many researchers are realized,
the day will soon come when virtually any product from pocket knives to
eyeglasses can be coated inexpensively with a brilliant, protective film of
diamond.

It seems paradoxical that diamond, the quintessential deep-earth
mineral could form in a near vacuum, but ultimately that may be the
cheapest, fastest, and most versatile of all diamond-making strategies.
At the earth's surface, clusters of carbon atoms usually adopt a pattern
with three nearest neighbors – the structure of graphite. But in the high-
vacuum vicinity of stars, hot carbon atoms often wind up linking to four
carbon neighbors, thus forming an interstellar diamond dust. What has
happened in space for billions of years can now be duplicated in a
variety of ways in the laboratory.

The key to the new family of diamond-making technologies, known
collectively as chemical vapor deposition or CVD, lies in forming a gas
of single, isolated carbon atoms (usually by heating carbon to very high
temperatures), and then fooling those atoms into adopting the "wrong"
structure when they cool. In successful CVD, the carbon vapor con-
denses to a crystalline layer so that each atom has four, rather than the
usual three, nearest neighbors – the distinctive structure of diamond.

This strategy works because hot, vaporized atoms have far more energy than atoms in either diamond or graphite. The carbon atoms' situation is analogous to a field of boulders precariously perched on the sides of a steep-walled valley. Boulders, like hot carbon atoms, are poised to attain a lower energy state. Eventually the boulders will fall, but they don't have to tumble all the way to the very bottom of the valley to come to rest – they may find a depression or trough part way down and stop there. Similarly, carbon atoms will condense from vapor to a lower-energy crystal, but they don't have to condense as stable graphite – the lowest energy structure. By providing a diamond seed or other appropriate substrate, carbon vapor will sometimes condense more readily as metastable diamond.

This synthesis strategy is an old idea, quite familiar to chemists, who often play such tricks on atoms. Indeed, a big part of the chemist's job is to learn how to synthesize such metastable compounds. Almost everything you buy at the grocery store, hardware store, or pharmacy, for example, is made in large part of metastable chemicals. Even before World War I some scientists had suggested that diamond, similarly, might be grown under conditions at which graphite normally would form. A few tentative low-pressure experiments were attempted in Europe early in the twentieth century, yet none of these pre-1950 efforts was sustained or, evidently, successful.[2] Given the potential uses of diamond coatings, it was only a matter of time before corporate and government labs gave it a try.

* * *

Low-pressure synthesis of diamond, though seemingly a long shot compared to the high-pressure route, is potentially a bench-top technology that doesn't require expensive presses or processing, and thus was well worth the effort. The first concerted attempts to make diamond at low pressure took place shortly after World War II, when research programs were established in the United States and the Soviet Union.

Two rival American corporate projects were initiated, one at the Linde Air Products Division of Union Carbide in Tonawanda, New York, and the other at General Electric's research facility in Schenectady, New York. The two groups adopted similar approaches, trying to grow diamonds by vaporizing hydrocarbons such as natural gas (CH_4) or acetylene (C_2H_2) in a vacuum. The less successful General

Electric effort, conducted primarily by researchers David Turnbull, Richard Oriani, and William A. Rocco, began in 1950 almost simultaneously with GE's Project Superpressure, and it continued with strong corporate backing but without any appreciable success until 1957, when the first public sales of high-pressure synthetic abrasives obviated the need for the vacuum process.[3]

Union Carbide scientist William G. Eversole, though little recognized at the time, was more successful. He spent the better part of the 1950s trying to grow diamond, and became the first to discover a reproduceable method to deposit new layers of carbon atoms on hot diamond surfaces. Eversole's laboratory records indicate that he began studies in 1949, and achieved the first unambiguous diamond growth between November 26, 1952, and January 7, 1953 – two years earlier than General Electric's high-pressure synthesis. In additional experiments conducted throughout 1953 he duplicated the process and confirmed the formation of diamond. Two United States and one Canadian patents, submitted in 1958, evidently represent the only published record of this feat, but William Eversole is now generally acknowledged to have devised the first reproducible method for low-pressure diamond synthesis, and his work formed a solid foundation for subsequent advances.[4]

Eversole's somewhat tedious process required repeated cycles of vacuum deposition of carbon atoms, followed by high-pressure cleaning to remove unwanted graphite. He first placed seed diamond particles in a vacuum chamber and exposed them to a hot, carbon-rich gas such as carbon monoxide (CO) or methane (CH_4). Ever so slowly, a thin layer of new carbon atoms would coat the diamond seeds. Unfortunately, a significant fraction of those carbon atoms always seemed to adopt the undesirable black graphite configuration. After a few hours, Eversole would have to remove the diamond seeds from the vacuum chamber, place them in an autoclave, and subject them to an intense cleaning cycle by exposure to hot (1000°C), pressurized (50 atmospheres) hydrogen gas, which reacts with graphite and removes it much more quickly than diamond. These two steps, repeated over and over again, eventually led to a small but measurable growth of the diamond seeds. His procedure, which under the best circumstances deposited a paltry ten-millionth of an inch of new diamond per hour, received the first patents for a low-pressure method of diamond synthesis in 1962.

William Eversole's discovery, anticipated by decades of theory and experiment, marked a major advance in diamond making. Even so, the method was too slow to become commercially viable and it came at a time when brute force high-pressure methods seemed to be nature's best path to making diamonds. Consequently, the Union Carbide process generated no public excitement and received relatively little notice from the scientific community at large.

Eversole's technique was studied with great interest, however, by the small community of scientists intrigued with the possibilities of CVD diamond growth, and his discovery provided a strong impetus for at least two intensive efforts – one in the Soviet Union at the Institute of Physical Sciences in Moscow, the other in the United States at Case Western Reserve University in Cleveland, Ohio. These programs, building on Eversole's work, achieved new successes independently and nearly simultaneously in the late-1960s.

* * *

Soviet researcher Boris V. Derjaguin and colleagues in the surface phenomena group at the Institute of Physical Sciences began their CVD diamond studies in the 1950s, when they independently followed a tack similar to that of Eversole.[5] This dedicated research program, which only gradually became known to American scientists, lasted for the better part of four decades, engaged several dozen scientists, and resulted in more than 100 publications on low-pressure diamond synthesis through the mid-1990s.

The first Soviet experiments, recorded in a patent application by Derjaguin and a student, Boris V. Spitsyn, submitted in 1956 (though not granted until 1980), employed heated carbon tetrabromide (CBr_4), carbon tetraiodide (CI_4), or carbon tetrachloride (CCl_4) as the source of hot carbon atoms.[6] While it is not evident that CVD diamonds were made in any of these early experiments, the Soviet scientists gained valuable experience in vapor deposition techniques and diamond analysis.

Public release of the Eversole patents in 1962 evidently provided new direction and impetus to their effort, for the Soviet group's subsequent advances were rapid. Dimitri V. Fedoseev joined the team in 1965 and, together with Derjaguin, they tackled several critical problems that blocked commercial development of CVD diamond technologies. These problems included the unwanted coprecipitation of graphite, the unacceptably slow rates of diamond growth, and the burdensome necessity of diamond seed crystals. They tried numerous variants of the Eversole process: different sources of carbon atoms (including electrically vaporized graphite and complex organic molecules), an oxygen atmosphere to remove unwanted graphite, a rapid sequences of hot light pulses to prepare the seed diamond surface, the introduction of hot hydrogen into the reaction chamber, and many others.[7] Theirs was classic empirical science – if you try enough different ideas, eventually something is bound to work.

In the late 1960s, B. V. Spitsyn noticed that when hydrogen gas was heated to 1000°C or more along with natural gas in the vacuum chamber some diamond crystals in the cooler regions of the chamber grew as much as a thousand times faster than in other experiments – a respectable ten thousandth of an inch per hour, with much less graphite to boot. Though the mechanisms of this process were not understood until several years later, the use of a hydrogen atmosphere became a focal point of CVD research. In 1971 Derjaguin and Fedoseev revealed their diamond-making success in a book, in which they published spectacular photographs of a vapor-deposited layer of faceted diamonds.[8]

In a rival United States program, chemist John C. Angus and his colleagues Herbert A. Will and Wayne S. Stanko of Case Western Reserve University tackled CVD diamond synthesis starting with the Eversole process. Initially, they were content to duplicate and confirm the Union Carbide results: they, too, employed a discontinuous cyclic process, with alternate eight-hour growth and seven-hour cleaning steps. Angus and coworkers meticulously cleaned natural diamond powder in acid and sealed it in a vacuum. Growth of new diamond – as much as a twenty-four percent increase in weight – occurred when natural gas at a pressure of less than a thousandth of an atmosphere was heated to 1050°C and passed over the diamond powder. Their report was submitted to *Journal of Applied Physics* in November, 1967, and appeared in print half a year later with the title "Growth of diamond seed crystals by vapor deposition."[9] The three authors, Angus, Will, and Stanko, gave ample credit to Eversole's earlier research, which they had successfully reproduced. They emphasized, "it is surprising that this remarkable work has received so little attention."

The Case Western Reserve group experimented with a variety of mixtures of natural gas (CH_4) and hydrogen (H_2), and they found, as had

Derjaguin and colleagues, that the presence of hydrogen at higher temperatures greatly enhances diamond growth. At the time no one was quite sure how or why the hydrogen made a difference, but a shrewd guess by John Angus seems to have marked a pivotal point in the history of CVD diamond. Near the end of his 1971 lecture on CVD processes at a meeting on diamond synthesis in Kiev, Angus suggested that at the high temperatures of their experiments molecular hydrogen gas (with two hydrogen atoms linked together) breaks apart into atomic hydrogen (single hydrogen atoms), and that these isolated atoms may play a key role in diamond deposition.[10] According to one current hypothesis, hydrogen atoms easily attach to exposed carbon atoms on the growing diamond surface, thus maintaining the tetrahedral diamond structure until new carbon atoms can be added. In the absence of hydrogen, so the theory goes, some carbon atoms revert to the three-coordination of graphite, thus disrupting the regular diamond structure. Whatever its role might be, atomic hydrogen proved the key to new advances. Following the Kiev meeting, the Soviet CVD research team made dramatic strides by pursuing diamond growth in the presence of atomic hydrogen. Derjaguin's group was suddenly restricted from publishing any CVD details from 1971 to 1976, and no mention of atomic hydrogen appears in any Soviet publication until the late 1970s. It appears, however, that by 1972 the Moscow group had learned to use a 95% hydrogen, 5% natural gas mixture to achieve a continuous diamond deposition technique (i.e., without an extra graphite cleaning cycle). Not only were they able to deposit diamond at an impressive rate of a thousandth of an inch per day, but they also discovered that their diamond films could be deposited on non-diamond substrates such as gold and copper.[11] When the Soviets first announced these diamond-growing successes in 1976 they purposefully withheld important details of the process. Consequently, few Western researchers took the claims seriously. This understandable skepticism may have been heightened by an embarrassing incident of a few years earlier, when Derjaguin and his colleagues first announced, and then retracted, the discovery of "polywater," a presumed polymerized form of water.[12] In any event, the Soviet results were soon eclipsed by dramatic new findings in Japan.

Following the Kiev meeting, Japanese researchers also took up the CVD diamond challenge. Seiichiro Matsumoto, Nobuo Sekata, and their colleagues at the National Institute for Research in Inorganic Materials (NIRIM) in Tsukuba, Japan, began an intensive study of the role of atomic hydrogen, first by duplicating American and Russian work, and then setting their own course towards commercialization.[13]

The NIRIM team greatly increased the reliability and convenience of the CVD process by generating hot carbon and hydrogen atoms with much simpler techniques than their Russian and American predecessors. Rather than use electric arcs or torches, they relied on a hot tungsten filament (as used in ordinary incandescent light bulbs) or microwaves to produce a vapor of carbon and hydrogen atoms. Once these techniques were established, Japanese workers devoted much of their efforts to depositing a durable diamond film on other materials, such as metals, glass, and plastic.[14]

The most flamboyant Japanese experiments, reminiscent of Robert Wentorf's peanut-butter-to-diamond exploits, were performed by Y. Hirose, who demonstrated the conversion of a wide variety of carbon-based materials, including saki, to diamond by CVD techniques.[15] In a subsequent demonstration, Hirose used an ordinary oxyacetylene torch to grow clear diamond films.[16] Robert De Vries quips that these advances "permit all of us to make diamonds in the garage or in front of a science class and to properly toast the success at the same time."[17]

Unlike their Russian counterparts, the NIRIM team did not hesitate to publish details of their new methods in the technical literature. Soon, U.S. and European laboratories were inspired to re-enter the field, and an intense international research effort continues to this day.

* * *

In spite of the pioneering research by William Eversole and John Angus in the 1950s and 60s, relatively little follow up occurred in North America, and leadership in CVD diamond technology had clearly shifted to Russia and Japan by the mid 1970s. Gradually, in the 1980s, that pattern began to reverse. Spurred by Japanese success, General Electric reactivated their moribund CVD project in 1984, under the leadership of Thomas R. Anthony and Robert C. DeVries. Norton, Du Pont, and Air Products (under the name Diamondex) also began CVD research, while a new company, Crystallume in Menlo Park, California, has devoted itself exclusively to CVD diamond manufacturing. Several universities, including North Carolina State, Penn State, and MIT, also commenced diamond film research in the 1980s.

An important impetus to U.S. research efforts was supplied by

Rustum Roy, Professor of Materials Science at the Pennsylvania State University. Roy is uncompromising, outspoken, and often controversial about science policy, but he is also remarkably astute in matters related to materials research. When Roy talks, most of his colleagues listen very carefully. In 1985, Roy visited CVD labs in both Russia and Japan, and he realized the tremendous potential of the rapidly evolving CVD technologies – a field that the United States had almost completely abandoned. Upon his return to Pennsylvania, he assumed a leadership role in the organization of a national consortium of government, corporate, and university researchers to share data and establish research priorities in CVD diamond and other materials frontiers.

One top priority identified by the materials consortium was the development of diamond semiconductors, which represent perhaps the most far reaching and profitable potential use for CVD diamonds.[18] Roy and others realized that the vapor deposition techniques widely used to manufacture silicon-based semiconductor devices by building up computer chips atom-by-atom are also potentially well suited to making CVD diamonds.

Semiconductors are the defining materials of our electronic age – they are the stuff of transistors, diodes, and integrated circuits. Today, most semiconductor devices are formed from silicon – a cheap material, easy to process, but quite susceptible to damage by heat, radiation, or other trauma. Some day diamond films, which are tough, durable, and resistant to heat damage, may become the ultimate semiconductor material. Major roadblocks remain, however, and these barriers are consuming vast amounts of research time and energy in today's high-tech laboratories.

Every electrical device, whether run by flashlight batteries, a gasoline generator, or electric power from the wall, functions by the artful control of electrons. To use electricity we must first learn to control the movement of these electrons – a task that depends on an impressive variety of high-tech materials. The two most obvious types of electronic age materials are electrical conductors and electrical insulators. We need conducting materials like copper wire in which electrons flow easily, so power can move from the electric plant to homes without dissipating. At the opposite extreme of the resistance scale we need insulators like plastic and glass that ordinarily don't let any electrons flow, so they protect us from dangerous electric currents.

If insulators and conductors were all we had to work with, the electronics field would be rather dull. Fortunately, there is another class of materials – semiconductors – that conduct electrons, but not very well. The behavior of electrons in a semiconductor is somewhat analogous to cars in an immense, crowded parking lot. Each car idles as its driver hopes to move to a better parking space. When every space is filled, nobody moves. Pure silicon, with just enough electrons to fill all the electronic parking places, is a poor electrical conductor.

Add just a few extra cars to the lot and the situation looks very different. Those cars without a parking spot rush around here and there, looking for a place to stop. In a silicon crystal scientists achieve the same effect by removing a few silicon atoms (just one in ten million will do) and replacing them with phosphorus atoms, each of which has an extra electron. Suddenly there are a few mobile electrons that can't be held and, bingo, you have an n-type semiconductor (because those extra electrons carry a negative charge).

A similar phenomenon occurs in the parking lot when there are a few empty parking spaces: for every hole some eager driver dashes out of his space to grab what he thinks might be a better one. The vacated space is quickly filled by another car, and so on. In a silicon crystal you can mimic that situation by substituting aluminum atoms once out of every few million silicons. Aluminum has one too few electrons, and so contributes the "holes" that inspire electrons to move around. The result is called a p-type semiconductor (because the holes are, in effect, positively charged).

Taken by themselves, n- and p-type semiconductors are just poor conductors but when pieces of the two types are combined into a single device something remarkable happens to the electrons. Electrons easily travel from n to p, because negatively charged electrons seek the positive holes, but if you try to push current the other way, from p to n, nothing happens. The composite of two kinds of semiconductor behaves very differently from simple insulators and conductors. The junction of an n and p semiconductor – technicians call it a diode – acts like a one-way street for electrons. Certain combinations of three semiconductors, npn or pnp, are called transistors and they may be used to amplify, detect, or switch electric signals. In modern microelectronics large numbers of n and p semiconducting regions form the complex integrated circuits, or microchips, that lie at the heart of all electronic gear.

* * *

Diamond has the potential to become the most reliable, efficient, long-lasting, and durable semiconductor known.[19] Of course, pure diamond won't do. Like pure silicon, with all its electronic parking spaces filled up, diamond is a poor conductor of electricity.

The problem of a *p*-type diamond semiconductor was easily solved. Scientists had long known that boron (an element with one fewer electrons than carbon), turned insulating diamond into a beautiful, efficient, deep-blue semiconductor. Indeed, John Angus and his students produced boron-doped diamond layers as early as the 1960s in their efforts to prove that their thin films were, in fact, diamond (since other carbon-rich compounds don't turn blue with the addition of boron).[20]

The critical missing link in the diamond semiconductor story is the lack of a comparable, easy-to-make *n*-type diamond semiconductor. Nitrogen, with one more electron than carbon, is the obvious first choice, but nitrogen doesn't release its extra electron in the diamond structure. Nitrogen-bearing diamond is still an excellent electrical insulator. Another candidate is phosphorus, which also has one more electron than carbon in its outer shell. But phosphorus is a bit too large to fit comfortably in a diamond host; it's difficult to coax enough phosphorus atoms into diamond to make a respectable *n*-type semiconductor.

Until one is found, there can be no diamond integrated circuits, but keep watching the news. Hundreds of researchers are working on the problem, and the age of revolutionary diamond semiconductors manufactured by CVD technology could be close at hand.

* * *

Most modern research, aimed directly at commercialization of CVD diamond, confronts several serious challenges. Many commercial applications will require extremely rapid continuous growth rates of at least a tenth of an inch per day, but at those deposition rates all current methods result in considerable graphite impurities. Crystal clear diamond coatings generally require growth ten to a hundred times slower. Thus, researchers continue to try a variety of methods, varying the carbon and hydrogen source, using different heating techniques, modifying the substrate material, and exploring a range of temperatures and pressures.

Materials scientists also are striving to produce large single crystal coatings, in which the arrangement of carbon atoms adopts only one orientation. These single-crystal layers are important for producing high-quality electronic devices. Most diamond films, however, are polycrystalline, with a mass of interlocked tiny diamonds oriented every which way. Scientists must find ways to prepare special substrates to coax all the carbon atoms into an orderly array.

Many of the most intriguing diamond film applications involve coating everyday products – permanently sharp cutlery, scratch-proof glasses and watch crystals, stickless kitchenware, durable machine tools, and distortion-free speaker components, are just a few of the potential uses. The problem is that the most efficient CVD methods require a substrate heated to more than 850°C – much too hot for plastics, glass, aluminum, or common household materials. Hundreds of researchers are tackling the specialized conditions necessary to apply diamond coatings to such everyday substances.[21]

Still other investigators are attempting to scale up CVD operations to coat large areas and complexly shaped machine tools. Clear diamond components up to several inches across, such as optical windows and rocket nose cones, have been manufactured (at considerable expense). But are there any limits to the size of a diamond object produced by CVD?

Some materials scientists question whether CVD diamond will ever become important commercially. They argue that vapor-deposited diamond is expensive to produce, and that its properties, while unique, do not yet satisfy any obvious critical need in the world of materials. Other more optimistic researchers, however, envision wonderful new uses for CVD diamonds.

Most CVD technology is now limited to depositing a thin diamond layer on an existing substrate, but what is difficult now may become routine in the decades to come. As our ability to deposit diamond increases, we may eventually learn to grow large, perfect diamond crystals of any size and shape. Imagine an era of solid diamond manufacturing, when we could create any object atom-by-atom, out of solid diamond. Diamond semiconductor devices, a durable alternative to silicon, would be foremost on many scientists minds, but solid-diamond technology could lead to the ultimate ball bearings, lightweight permanent artificial joints, and even solid diamond household utensils. Sculptors would possess a new medium of unrivaled brilliance, while artisans could shape solid diamond jewelry of breathtaking beauty.

Thanks to CVD techniques, the day may come when diamond, the most prized of all gems, becomes as familiar in everyday life as plastic or glass.

Notes

1. John C. Angus, as quoted in: Ivan Amato, "Diamond fever." *Science News* **138**, 72–74 (1990), p.72.

2. Historical and technical details related to metastable diamond synthesis were provided through letters and interviews with Robert C. DeVries, Rustum Roy, and Russel Seitz. Early efforts to produce diamonds by chemical vapor deposition are reviewed by: Robert C. DeVries, "The science behind the synthesis of diamond. Part I: Equilibrium conditions. Part II: Nonequilibrium conditions." *Journal of Materials Education* **13**, 387–440 (1991); R.C. DeVries, A. Badzian, and R. Roy, "Diamond synthesis: The Russian connection." *Materials Research Society Bulletin* **21**, 65–75 (1996), p.70–71.

An overview of CVD diamond science and technology is provided by: National Materials Research Board, *Status and Applications of Diamond and Diamond-Like Materials: An Emerging Technology.* (Washington: National Academy Press, xii, 98 pp. 1990). Many aspects of the CVD process are also reviewed in: Karl E. Spear and John P. Dismukes (editors), *Synthetic Diamonds: Emerging CVD Science and Technology.* (New York: Wiley, 663 pp. 1994). Review articles that describe the principles and chronicle the history of metastable (low-pressure) diamond synthesis include: John C. Angus and Cliff C. Hayman, "Low-pressure, metastable growth of diamond and diamondlike phases." *Science* **241**, 913–921 (1988); Peter K. Bachman and Russell Messier, "Emerging technology of diamond thin films." *Chemical and Engineering News* **67**, 24–39 (1989); Walter A. Yarbrough and Russell Messier, "Current issues and problems in the chemical vapor deposition of diamond." *Science* **247**, 688–696 (1990); Ivan Amato, "Diamond fever." *Science News* **138**, 72–74 (1990); F. G. Celii and J. E. Butler, "Diamond chemical vapor deposition." *Annual Review of Physical Chemistry* **42**, 643–684 (1991); J. C. Angus, Y. Wang, and M. Sunkara, "Metastable growth of diamond and diamond-like phases." *Annual Review of Materials Science* **21**, 221–248 (1991); Claus-Peter Klages, "Metastable diamond synthesis – principles and applications." *European Journal of Mineralogy* **7**, 767–774 (1995).

3. R. A. Oriani and W. A. Rocco, "Attempts to grow diamond under metastable conditions." GE Memo No. MA-36, Class IV, August, 1957.

4. W. G. Eversole, "Synthesis of diamond by deposition on seed crystals." U.S. Patent No. 3 030 187 and 3,030,188, filed July 23, 1958; issued April 17, 1962. See also Canadian Patent No. 628,567 of the same name. Details of Eversole's work are related in: J. C. Angus and C. C. Hayman, "Low-pressure, metastable growth of diamond and diamondlike phases." *Science* **241**, 913–921 (1988).

5. R.C. DeVries, A. Badzian, and R. Roy, "Diamond synthesis: The Russian connection." *Materials Research Society Bulletin* 21, 65–75 (1996).

6. B. V. Spitsyn and B. V. Derjaguin, USSR Patent No. 339,134, issued May 5, 1980.

7. Some of the Soviet CVD experiments are reported in: B. V. Derjaguin, D. V. Fedoseev, V. M. Lukyanovich, B. V. Spitsyn, V. A. Ryabov, and A. V. Lavrentyev, "Filamentary diamond crystals." *Journal of Crystal Growth* 2, 380–384 (1968); and B. V. Derjaguin and D. V. Fedoseev, "Physico-chemical synthesis of diamond in metastable range." *Carbon* 11, 299–308 (1973).

8. B. V. Derjaguin, D. V. Fedoseev, V. N. Bakul, V. A. Ryabov, B. V. Spitsyn, Y. I. Nikitin, A. V. Bochko, V. P. Varnin, A. V. Laurent'ev, and V. L. Primatchuk, *Physicochemical Synthesis of Diamond from Gas.* (Kiev: Tecknika, 1971). See also: B. V. Derjaguin and D. V. Fedoseev, "The synthesis of diamond at low pressures." *Scientific American* 233, 102–109 (1975).

9. John C. Angus, Herbert A. Will, and Wayne S. Stanko, "Growth of diamond seed crystals by vapor deposition." *Journal of Applied Physics* 39, 2915–2922 (1968).

10. The Case Western Reserve paper, presented at the International Conference on Synthetic Diamonds in Kiev, USSR (February 14–18, 1971) was: J. C. Angus, D. J. Poferl, N. C. Gardner, S. Chauhan, T. J. Dyble, and P. Sung, "Growth of diamond and semiconducting diamond at subatmospheric pressures."

The remarks by Angus may have been prompted, in part, by a little-known previous study of Henry J. Hibshman, who received U.S. Patent 3,371,996 for "Diamond growth process" on March 5, 1968. That patent, filed January 20, 1964, states "in the presence of platinum the hydrogen is converted to atomic hydrogen which destroys the nucleii of graphite or other black forms of carbon as rapidly as such nucleii are formed." While some of Hibshman's patents were acquired by Esso Research and Engineering Company, Linden, New Jersey, there is no indication that this patent was ever commercially implemented.

11. B. V. Spitsyn, L. L. Bouilov, A. A. Klochkov, A. E. Gorodetskii, and A. V. Smolyanikov, "Diamond crystal synthesis on nondiamond substrates." *Doklady Akademie Nauk SSSR* 231, 333–335 (1976); a translation appears in *Soviet Physics – Doklady* 21, 676–677 (1976).

12. For a more complete discussion of the polywater story see Chapter 13 and reference therein.

13. S. Matsumoto, N. Setaka, "Consolidation of diamond powders by thermal decomposition of methane and benzene." *Journal of Materials Science Letters* 15, 1333–1336 (1980).

14. S. Matsumoto, Y. Sato, M. Tsutsumi, and N. Setaka, "Growth of diamond particles from methane-hydrogen gas." *Journal of Materials Science* 17, 3106–3112 (1982); and M. Kamo, Y. Sato, S. Matsumoto, and N. Sekata, "Diamond synthesis from a gas phase in microwave plasma." *Journal of Crystal Growth* 62, 642–644 (1983).

15. Y. Hirose, "Synthesis of diamond thin films by thermal CVD using organic compounds." *Japanese Journal of Applied Physics* Part 2, vol. 25, L519–L521 (1986).

16. Y. Hirose, "Synthesis of diamond using a combustion flame in the atmosphere," in S. Saito, O. Fukunaga, and M. Yoshikawa (editors), *Science and Technology of New Diamond* (Tokyo: Terra Scientific Publishing Co, 1990), pp. 51–54.

17. R.C. DeVries, A. Badzian, and R. Roy, "Diamond synthesis: The Russian connection." *Materials Research Society Bulletin* 21, 65–75 (1996), p.73.

18. National Materials Advisory Board, *Status and Applications of Diamond and Diamond-Like Materials: An Emerging Technology.* (Washington: National Research Council, xii, 98 pp. 1990).

19. Michael W. Gels and John C. Angus, "Diamond film semiconductors." *Scientific American* **267**, 84–89 (October, 1992).

20. D. J. Poferl, N. C. Gardner, and J. C. Angus, "Growth of boron-doped diamond seed crystals by vapor deposition." *Journal of Applied Physics* **44**, 1428–1434 (1973).

21. Numerous recent articles describe techniques for applying diamond coatings by vapor deposition. See, for example: J. L. Robertson, S. C. Moss, Y. Lifshitz, S. R. Kasi, J. W. Rabalais, G. D. Lempert, and E. Rapoport, "Epitaxial growth of diamond films on Si(III) at room temperature by mass-selected low-energy C^+ beams." *Science* **243**, 1047–1050 (1989); M. D. Drory and J. W. Hutchinson, "Diamond coating of titanium alloys." *Science* **263**, 1753–1755 (1994); J. Narayan, V. P. Godbole, and C. W. White, "Laser method for synthesis and processing of continuous diamond films on nondiamond surfaces." *Science* **252**, 416–418 (1991).

CHAPTER 13

THE DIAMOND BREAKERS

You can observe a lot by watching.
YOGI BERRA

F OR THE BETTER PART OF A CENTURY scientists relied on the earth to learn how diamonds were made. Today, they rely on diamonds to learn how the earth was made.

Diamond's utility in modern research stems in large measure from its strength and hardness, but it possesses another extraordinary property. Diamond is unique in its transparency to all kinds of light.

All of us, every moment of our lives, are surrounded by a surging sea of electromagnetic radiation. Light – energy in the form of waves that travel 186 000 miles per second – pervades every nook and cranny of the universe. Our eyes sense only a tiny fraction of this energy; there is more – much more – out there.

More than a century ago James Clerk Maxwell predicted the existence of invisible kinds of light. He realized that waves of light could be any length at all, from the diameter of an atom's nucleus to the size of stars, the same way ocean waves can range from microscopic ripples to globe-spanning swells. Like sailors clinging to a bobbing boat, who can only sense a small range of the ocean's waves, our eyes see only a small portion of the electromagnetic spectrum – the part where the waves are about a ten-thousandth of an inch long. Red, orange, yellow, green, blue, violet – the colors of the spectrum are nothing more than thin slices of the vast radiation continuum.

The discovery of the other kinds of light – radio waves, microwaves, infrared radiation, ultraviolet light, x-rays, and gamma rays – propelled science and technology into the twentieth century. Scientists and engineers quickly learned that each kind of light energy can be absorbed, scattered, or transmitted to advantage.

In order to manipulate light you first have to know how it interacts with matter. With so many materials and such a wide range of light, it's a huge job that continues to consume the research lives of tens of thousands of scientists, who call themselves by lots of fancy names like spectroscopists, crystallographers, or radio-astronomers.

Research on matter and light has revealed an extraordinary fact about diamond. Of all known solids, nothing is more transparent to a wider range of light.[1] Not only can you look right through a diamond with visible light, but the gemstone is also largely transparent to radio waves, microwaves, x-rays, gamma rays, and most kinds of ultraviolet and infrared light as well. Diamond is the ultimate window. High-pressure scientists, armed with knowledge of diamond's unique transparency, didn't take long to devise a completely new use for diamonds. For them, the gemstone provided an unparalleled way to measure volume.

Scientists have observed many unexpected phenomena at high pressure, but one behavior is always true – squeeze on anything and it will get smaller. Whenever you apply force to a pencil, push a button, or blot a paper, the atoms under stress will, for a time, move a little closer together. Scientists quantify this effect with graphs of pressure versus volume – the meat and potatoes of high-pressure research – which identify the abrupt atomic rearrangements called "phase transitions," such as the transformation of graphite to diamond, or water to ice. High-pressure workers are always looking for new phase transitions, and they know that pressure-volume data are often the best tools for spotting them. They also ask lots of questions about their compressed samples: "Which atoms get closer together?" "How much closer do they get?" "Do some kinds of atoms compress more than others?" Such questions about atoms and how they are arranged are usually the intellectual turf of the crystallographers.

Crystallographers focus x-ray beams at tiny crystals, and study how the energy scatters. Each type of crystal yields a distinctive pattern of scattered x-rays, and the crystallographer's job is to work backward – measure the x-ray pattern and deduce the atomic structure. By the close of World War II x-ray crystallography had become routine, and most common atomic structures had been thoroughly described. Enterprising researchers, always seeking fresh turf, looked to the intriguing problem of atomic structures at high pressure. What they needed was a material strong enough to hold a pressurized sample, and transparent enough to zap the sample with x-rays. The answer was diamond.

* * *

In 1946, many of the tens of thousands of physicists, chemists, engineers, and technicians who helped make the atomic bomb during the epic Manhattan Project were looking for new opportunities. Scientists at the innocuous sounding "Institute of Metals" at the University of Chicago were among those who spent the War working on the bomb, but the end of hostilities didn't end their involvement in weapons research. In 1947 they received a modest grant from the Office of Naval Research to study the behavior of metals and other materials at high pressure.[2] Their work was hardly arcane – you can't build submarines or calculate the destructive effects of explosives without knowing how metals behave under pressure.

Physicist Andrew W. Lawson, like so many of his colleagues, spent much of the War years on atomic bomb research at Chicago. He proved an able leader and, in spite of his youth, was named head of the Chicago high-pressure project in the spring of 1947. Within a few weeks Lawson took the logical first step – a pilgrimage to Harvard to confer with Percy Bridgman. In typically generous fashion Bridgman gave Lawson the blueprints for his press and pump, and provided all sorts of advice on laboratory procedures. Inevitably, their pressure conversation turned to the question of volume.

The classic problem for high-pressure researchers is how to measure the volume of a sample crushed inside a massive metal device. Bridgman and others did it by the simple method of piston displacement: they measured the amount a piston moved in response to changes in the sample's volume. The technique worked reasonably well, but told him nothing about what the individual atoms were up to. Lawson and Bridgman knew that x-rays, which can probe a sample's atomic structure, could give them the answer. All they needed was a material that could sustain high pressure while transmitting x-rays.

X-rays are not absorbed by atoms themselves, but by the clouds of electrons that surround each atom's nucleus. One key to finding a material transparent to x-rays is to find atoms with only a few electrons.

Nature hasn't given us many workable choices. The lightest atoms, elements 1 and 2 (hydrogen and helium, with one and two electrons per atom, respectively), are both gases so they don't make very good pressure cells. Element 3, lithium, is an extremely soft and chemically reactive metal, so it won't work either. Which takes us to the fourth element, the metal beryllium, with four electrons per atom.

Beryllium is a shiny, silvery metal with a slight bluish sheen; it looks a lot like stainless steel at first glance. It's easy to machine, though potentially quite toxic to the machinist if precautions against beryllium dust aren't taken. The University of Chicago group built a pressure cell entirely out of the expensive metal and obtained reasonably good x-ray patterns at 15 000 atmospheres. After a few trials, however, it was clear that the metal cell would always be limited by the thick beryllium walls; there were simply too many atoms in the way. The cell absorbed most of the x-rays and scattered much of the rest to fog their x-ray film. Then Lawson and his crystallographic colleague Ting-Yuan Tang turned to diamond, formed from the six-electron element, carbon. They reasoned that the great strength of diamond would permit a much smaller pressure device with lower x-ray absorption.

These two scientists pioneered the use of diamonds in high-pressure x-ray research. Their rudimentary device, first assembled in 1949, consisted of a small diamond cut in half and then clamped together in a steel vise.[3] A hole no larger than the diameter of a pin was laboriously drilled through the diamond halves, parallel to their contact, to form a cylindrical sample chamber (fig. 50). Two short lengths of steel piano wire or tungsten carbide, inserted at opposite ends of the hole, compressed powdered samples to more than 20 000 atmospheres. Following the century-old tradition of high-pressure researchers familiar with explosions, Lawson and Tang dubbed their invention the split-diamond bomb.

In trial x-ray experiments on calcite (the common white mineral that makes most limestone and marble), they demonstrated the feasibility of taking x-ray pictures through diamond. Lawson and Tang inserted a powdered sample into the cylindrical hole and compressed it from both ends with wire. They focused an x-ray beam through the diamond onto the pressurized sample, and recorded the resulting diffraction with special x-ray film. The split diamond bomb proved that diamonds were an ideal window for high-pressure x-ray work, but the device failed to take full advantage of the gemstone's strength. By splitting the diamond

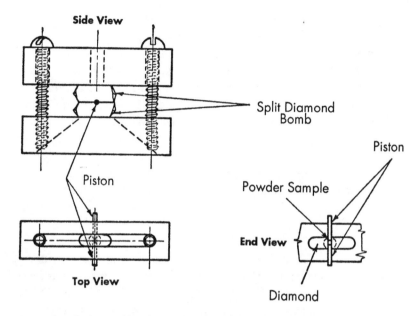

Fig. 50 The split-diamond bomb of Andrew Lawson and Ting-Yuan Tang consisted of a cleaved diamond held together by a screw-tightened metal clamp. A hole drilled through the diamond served as a sample chamber, while a stiff wire piston compressed the powdered sample from either side of the hole. (Courtesy of the American Institute of Physics.)

in two, the Chicago researchers weakened the cell and limited the maximum attainable pressure. As so often happens in science, an eager graduate student took the next step.

John C. Jamieson, who would later work with Paul De Carli to synthesize diamond by explosion, was a jovial, genial man who seemed to approach each new experimental challenge with cheerful optimism. Upon joining the University of Chicago high-pressure team in 1951 as a doctoral candidate he immediately set to work on the high-pressure calcite problem. Jamieson reasoned that a hole drilled through a single crystal of diamond would be stronger, and therefore confine higher pressures, than a halved diamond. He selected a three-carat gem, worth more than $50 000 in today's dollars, and took it to a Chicago company that specialized in drilling holes in diamond for pulling out wire of uniform diameter. Shortly before his death in 1983, Jamieson recalled his uncharacteristic impatience at having to wait eight months for the process to be completed. With a twinkle in his eye, he told of his phone calls every few weeks to track the progress of a $1/4$ inch-long hole through the hardest known material. Eventually, bit by bit, as diamond drill abraded diamond crystal, the hole was completed.

The new cell worked beautifully, achieving 30 000 atmospheres and allowing Jamieson to discover a previously unknown phase transition in calcite.[4] He also heated samples in the diamond bomb and thus mimicked the behavior of minerals in the hot, compressed interior of the earth. Jamieson's single-crystal cell could do things no other device could match, and other labs rushed to make one of their own. Yet, as clever as the miniature piston-cylinder was with its transparent diamond walls and metal pistons, it just missed the mark. The real diamond cell breakthrough took a brilliantly different approach.

* * *

Few scientists have had to resist greater temptations than Alvin Van Valkenburg and Charles Weir, who formed the high-pressure research team of the National Bureau of Standards (NBS) in Washington, DC.[5] It was there that Van Valkenburg and Weir were given unlimited access to fistfuls of gem diamonds.

Van Valkenburg came to NBS just after World War II, having studied geology, mineralogy, and x-ray crystallography. There he joined forces with Charles E. Weir, probably the only African-American to play a major part in the post-War development of high-pressure research.

Van Valkenburg's career had progressed easily, but Weir had struggled all his life against prejudice. His scientific credentials were impeccable: an undergraduate physics degree from Chicago and graduate work at Howard University and Caltech. The NBS newsletter referred to him euphemistically as "one of those rare native Washingtonians," but even in the community of supposedly objective physicists racism was pervasive. The restaurant at the National Bureau of Standards made it clear that he was not welcome, so Van Valkenburg and Weir usually ate sandwich lunches together in their office. Discrimination was even more blatant at many professional meetings, where Weir was forced to eat and sleep away from the conference center. In New York City he always ate at the coin-operated cafeterias, where everyone was treated the same.

Van Valkenburg and Weir formed a lasting friendship at the NBS, where they had one unusual advantage over other high-pressure workers. As government researchers, they enjoyed a steady supply of fine diamonds confiscated from would-be smugglers. There have always been gem smugglers, but the greatest diamond hoard came to America following Britain's decision to pull out of Palestine in 1948. Before escaping the danger zone, many people had converted their wealth into diamonds, and many of those stones were illegally carried into the United States. Customs agents in New York and Florida confiscated millions of dollars worth of smuggled stones, which the government refused to release onto the carefully regulated commercial market. Boxes full of these precious stones were first offered to the scientists at the National Bureau of Standards.

"When you do science you have to give credit to all who helped," Van Valkenburg reminisced. "The smugglers made my career." He never learned their names, though he was told that more than one went to jail. Thanks to them, the two NBS workers had thousands of diamonds at their disposal – valuable gems with no inventory, just sitting there for the taking. Most of the diamonds were small, weighing only a fraction of a carat, but there were scores of larger stones as well. The standout was a magnificent 7.5–carat cut gem of a particularly rare type that displayed transparency to infrared radiation.

Van Valkenburg and Weir handled their treasure casually. Scientists often keep messy offices, but Van Valkenburg and Weir were worse than most. The desks in their small shared office were always piled high with unread papers and unanswered mail. To make matters worse, Charlie Weir smoked incessantly and scattered ashes were everywhere. Amid this chaos, diamonds were just lying around, waiting to be studied.

Van Valkenburg recalled one diamond delivery – perhaps a million dollars' worth – that arrived by courier late one afternoon. It was too late to arrange for proper security, and the night watchman refused to take responsibility for the gems, so Van Valkenburg just stuck them in the back of his desk drawer, where they lay forgotten for several days. Given all the government lab regulations of the 1990s, it's difficult to imagine such lax supervision, but there were few controls on NBS scientists in the 1950s.

One time Van Valkenburg accidently dropped a small diamond – no more than a third of a carat but still worth a week's pay – on the dusty floor. In spite of a careful search, the gem seemed to have vanished; it turned up a week later in the far corner of the lab and was plopped back into its box. This cavalier attitude was prevalent throughout the NBS;

one by-product was the bureau's almost total lack of environmental concern. Smoke stacks of the NBS glass group spewed all sorts of toxic fumes – beryllium, arsenic, lead – right in the middle of a quiet residential Washington, D.C. neighborhood. No one said anything about it.

With so few regulations and so many diamonds, Van Valkenburg and Weir first thought about how to study the mineral itself. They knew that not all diamonds are alike. While the great majority absorb a broad range of invisible infrared radiation, perhaps one of every hundred natural gems is transparent to infrared wavelengths. Scientists could easily shine an infrared beam onto stones to sort diamonds into Type I (opaque to infrared) and Type II (transparent), but they didn't have a clue about what caused the difference. Some said it was a subtle structural difference, while others argued for an unknown impurity element.[6]

Van Valkenburg favored the impurity hypothesis, and he burned dozens of diamonds, analyzing the resulting gas and hoping to identify the foreign element. The mystery element wasn't boron, oxygen, aluminum, phosphorous, or sulfur. After incinerating dozens of gems, he gave up in frustration. (Nitrogen, an element Van Valkenburg couldn't detect with his apparatus, was eventually shown to be the culprit.)

As Alvin Van Valkenburg was busy cremating precious stones at the National Bureau of Standards, another government agency was vying for the diamond hoard. George Switzer, gem curator at the Smithsonian's Natural History Museum, was desperately trying to work a deal to acquire the Hope Diamond, the most notorious of all blue diamonds. Switzer thought a trade including the multimillion-dollar stockpile of smuggled diamonds might convince its owner, Harry Winston, to part with the fabulous jewel. Thus, for a time in the mid-1950s, there was considerable pressure on Van Valkenburg and Weir to produce something more useful from the stones than carbon dioxide. Although the Hope Diamond trade never came to pass (in 1958 Winston finally just gave the famous stone to the Smithsonian), the episode did cause the NBS scientists to look for better uses for cut diamonds.

Having burned numerous diamonds to no avail, Van Valkenburg and Weir discovered other ways to destroy the gems. They selected the magnificent 7.5–carat, emerald-cut, Type II gem – worth perhaps a quarter of a million dollars in today's retail market – for their copy of Jamieson's University of Chicago drilled diamond cell. If Jamieson got 30 000 atmospheres with a 3–carat stone, they reasoned, NBS should be able to go even higher with a 7.5–carat jewel. It took four solid

Fig. 51 Alvin Van Valkenburg, NBS research technician Elmer Bunting, and Charles Weir (from left to right) pose with their drilled-diamond cell. The device (insert, lower right) was made from a 7.5-carat gemstone that finally shattered at high pressure (R. M. Hazen).

months of drilling and consumed 16 carats of diamond dust to craft the tiny hole – 1/4–inch long and 1/64th-inch wide – but for a short while the NBS team had the world's most glamorous pressure cell (fig. 51).[7] Soon, however, they faced the friendly rivalry of other high-pressure groups. General Electric pressure experts, not wanting to be left out when it came to diamonds, constructed their own drilled cell, but with slightly tapered holes for an even tighter fit and even higher pressures. In the mid 1950s, drilled diamond cells at Chicago, NBS, and GE set records for the best x-ray and infrared spectra results at the highest pressures. Until 1957, that is, when the NBS team made their big mistake.

Pressure workers are greedy; they always want higher pressure. The NBS team, hoping to establish a new record, literally pushed their cell to the breaking point. With a sickening crack, the 7.5–carat gemstone shattered into four pieces. Too late they had learned a critical lesson: a diamond is incomparably strong when under compression, but place it under tension – push from the inside out – and it shatters.

Van Valkenburg and Weir were stunned and dismayed at their costly failure. What could they do? Drill another diamond, with the constant worry that it, too, might shatter? Abandon high-pressure research altogether? They debated their options for a few depressing days and made their decision: it was time to try something new.

* * *

The diamond-anvil cell, affectionately nicknamed the "DAC" in many high-pressure labs, was an idea waiting to happen. The diamond cell is so simple in design and construction that it could have been built and used centuries ago. Isaac Newton could have watched matter transform at thousands of atmospheres pressure. Thomas Edison could have discovered new materials with extraordinary properties. But it was not until 1959 that two research teams, independently and virtually simultaneously, hit upon the idea that transformed high-pressure research.

Perhaps the easiest way to enclose a sample and subject it to pressure is the piston-and-cylinder. Percy Bridgman, Loring Coes, George Kennedy, and many others adopted this approach in their large presses, where sturdy metals formed the cylindrical walls of the chamber and carefully machined steel or carbide pistons drove into the confined sample. The split and drilled diamond cells of the Chicago high-pressure lab incorporated the same strategy, with diamond acting as a strong and transparent cylindrical chamber, and stiff wire acting as the miniature pistons. It wasn't the best use of diamonds, but it was the logical first step to try.

Bridgman used another strategy – opposed anvils – to obtain the highest pressures of his career. That geometry was also adopted in one form or another by most of the early diamond makers – Tracy Hall's belt, Herb Strong's cone, even von Platen's split sphere incorporated pairs of anvils. With that history in mind, workers at both NBS and Chicago, stimulated by two very different experimental objectives but both committed to using transparent diamond windows, converged on a single, elegant solution. They devised diamond-anvil cells based on Percy Bridgman's opposed-anvil design.

John Jamieson created the new apparatus in the hope of achieving higher pressures for his x-ray studies. His drilled diamond cell, with a maximum pressure range of only about 30 000 atmospheres, could duplicate only the first fifty or sixty miles of the earth's outer layers – representing only a pitiful four percent of the solid earth's total volume. He realized that in the drilled diamond cell the diamond served only as a passive chamber, not the active element in applying pressure. Much better, he thought, to apply pressure using the diamonds as anvils, with the sample squeezed in between.

Thus the Chicago diamond-anvil cell was born. Jamieson took a bit of sample, put it on a diamond surface, placed a second diamond on top, stuck the whole works in a screw vise, and squeezed like crazy. X-rays reached the sample by passing through the sides of the relatively thin diamond anvils, parallel to the anvil faces, thus minimizing x-ray absorption (fig. 52). Jamieson and Lawson completed a prototype cell late in 1958, optimistic that 100 000 atmospheres would soon be within their reach.

The University of Chicago scientists used their diamond cell to produce excellent x-ray patterns of a new form of bismuth found at 35 000 atmospheres, a record for x-ray work, and they were only a little discouraged when one diamond cracked as they reached for higher pressures. The Chicago pressure cell represented a significant advance, and it received prominent exposure in the November, 1959, issue of *The Review of Scientific Instruments.*[8] But Jamieson and Lawson just missed

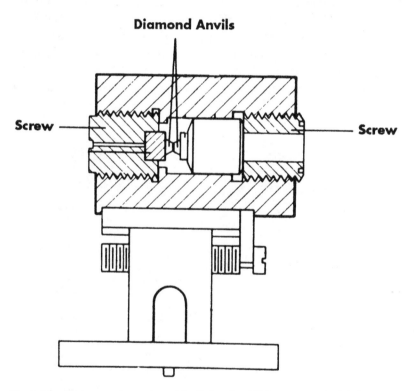

Fig. 52 John Jamieson and coworkers at the University of Chicago pressurized their samples by squeezing them between the faces of two diamond anvils, using two large screws to change the pressure. The entire assembly could be mounted onto an x-ray machine, which irradiated the sample from the side of the opposed anvils. (Courtesy of the American Institute of Physics.)

the full potential of the opposed-diamond topology. The NBS team
would have to show the world what diamonds could do.

* * *

The University of Chicago scientists had designed their diamond cell
for x-ray studies, but Van Valkenburg and Weir at the NBS attacked the
high-pressure problem from the different perspective of the infrared
spectroscopist. The Chicago team had passed an x-ray beam through
the side of the diamond anvils, analogous to the earlier drilled diamond
arrangement, but to do spectroscopy you have to see the sample, and to
see a high-pressure sample you have to be able to look *through* the
sample chamber (fig. 53). In late 1958 Van Valkenburg hit upon the
simple idea of cutting the tips off brilliant-cut gem diamonds and using
them as tiny transparent anvils. By leaving a hole in the steel anvil supports, he could look right through the anvils at the compressed sample
with no distortion or interference.

   Charles Weir, who counted machining skills among his credentials,

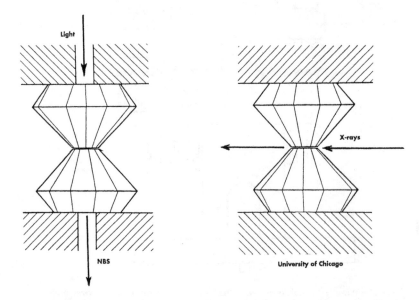

Fig. 53 Research groups at the University of Chicago and the National Bureau of Standards
adopted different strategies in their diamond-anvil research. NBS scientists directed their
infrared beams through the two diamonds perpendicular to their flat faces, while Chicago
workers passed an x-ray beam parallel to the diamond-anvil surfaces.

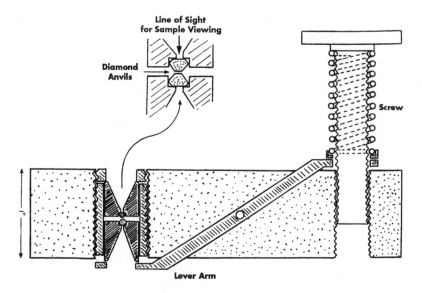

Fig. 54 The National Bureau of Standards's lever-arm diamond-anvil cell employed a pair of opposed diamond anvils mounted in steel pistons. The screw acted on the lever arm to change pressure. This device, designed for spectroscopic work, allowed scientists to look directly at their pressurized samples. (Courtesy of Eric Van Valkenburg.)

built the first NBS opposed-diamond-anvil device in 1958. He attached Type-II diamond anvils to two steel pistons, which fit into a precisely machined cylinder. The pistons were clamped together by a slick lever-arm arrangement that Van Valkenburg described as "little more than an elaborate nutcracker."[9] (fig. 54) The NBS group believed that exceedingly high pressures might be achieved in their machine with a simple turn of a screw.

The new diamond-anvil cell with its Type-II diamonds seemed ideal for infrared studies. The device fit snugly into their infrared machine, and Van Valkenburg and Weir were confident that obtaining infrared data would be a straightforward task. Poised to do great things, they decided to seek the help of a widely respected authority on infrared spectroscopy. They approached Prof. Ellis R. Lippincott at the University of Maryland's chemistry department, a gentleman scientist who spent a long and productive career teaching chemistry and contributed a great deal to our understanding of how matter absorbs and scatters light.

Lippincott was immediately drawn to the NBS proposal. He knew that pressure forces subtle shifts in atomic positions, and alters the way light rays of different colors travel through a solid. Apply pressure and a material's spectrum will change ever so slightly. Study high-pressure spectra, Lippincott thought, and you'll discover a lot about matter.

Lippincott, Van Valkenburg, and Weir performed their first infrared experiments on a pinhead-sized mound of potassium chloride. The NBS researchers, nervous with anticipation, piled their sample on one diamond face, crushed the white powder between vise-like anvils, and placed the cell in their spectrometer. No spectrum appeared. They shifted the pressure cell this way and that, refocused the beam, and adjusted its intensity, but to no avail. Van Valkenburg remembers wasting more than an hour in fruitless fiddling. In frustration, he removed the cell from the spectrometer and placed it under a microscope to see what was wrong.

At that moment, he became the first human to watch matter transform at thousands of atmospheres pressure. The potassium chloride around the edges of the diamond remained colorless, but at the center of the flattened disk-shaped sample, where pressure was much higher, was a circle of darker material – clearly a new high-pressure form of the chemical. One by one the NBS team looked in amazement at the microscopic phenomenon. "It seems odd," Van Valkenburg recalled, "but when we designed that diamond cell we never actually thought about *looking* at the sample." But look at samples they did. The Lippincott, Weir, Van Valkenburg patent for a "High-Pressure Optical Cell" ushered in a new era of high-pressure research. For the very first time researchers could not only measure spectra at a wide range of wavelengths – they could also actually watch an amazing variety of high-pressure phenomena at pressures greater than 100 000 atmospheres. Early photographs taken by the NBS group through their cell show striking phase transitions in materials at pressures corresponding to depths of more than 200 miles beneath the surface of the earth.

As powerful as their first diamond-anvil cell was, the NBS team had just gotten started. "We tried everything," Van Valkenburg remembers, "and we broke diamonds left and right." But they eventually achieved pressures of more than 300 000 atmospheres.

Van Valkenburg realized that the diamond-anvil cell was greatly limited because all of their samples were completely crushed between the diamond anvils. Single crystals, with their special directional properties, could not be studied, nor could liquids or gases be confined and probed with the original design. He devised an elegant solution – one he views in retrospect as almost trivial, though it radically changed the science of high-pressure.

Van Valkenburg observed that two flat diamond anvils aren't enough to confine a sample – they only form a contact surface. So he took a thin metal sheet of nickel alloy, drilled a hole in it, and used it as a gasket. By placing the metal gasket between the two anvils, he formed a tiny cylindrical sample chamber about the size of a sand grain (fig. 55).

Alvin Van Valkenburg saw wonderful things in his gasketed diamond cell. Simple water, squeezed to 10 000 atmospheres, formed rectangular crystals of a new kind of "ice" that no one had ever seen

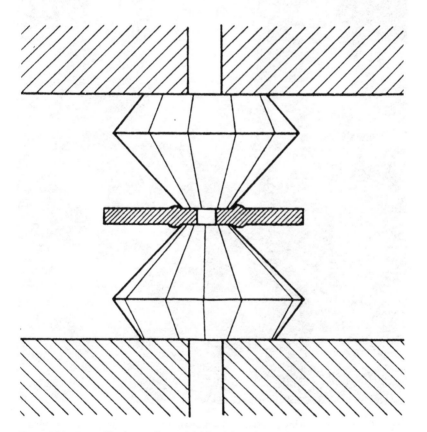

Fig. 55 Alvin Van Valkenburg enhanced the flexibility of the diamond-anvil cell by incorporating a metal gasket between the two diamonds. This gasketed cell enabled him to study fluids and uncrushed single crystals at high pressures.

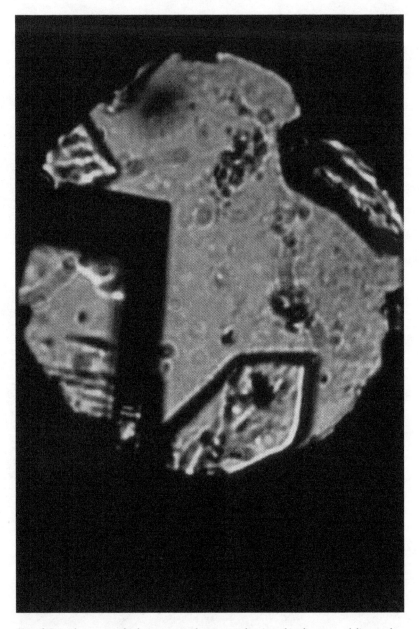

Fig. 56 Crystals grown in high-pressure solutions can be viewed in the opposed diamond anvil cell. Alvin Van Valkenburg took this photograph of calcium hydroxide, *circa* 1970 (R. M. Hazen).

before. Then, at 25 000 atmospheres, another new form with cube faces appeared in the cell. He saw everyday liquid chemicals like benzene and toluene form crystals at pressure, and dubbed the blade-like crystals of pressurized alcohol "gin-sickles."[10]

Metal gaskets also allowed the NBS researchers to study tiny single crystals without the grains being crushed between the diamonds. They placed crystals of calcite and other minerals in the gasket chamber, surrounded them with a liquid, and closed the cell up. Squeezing the diamonds together didn't smash the crystals; it just compressed the surrounding liquid, which in turn compressed the crystal equally on all sides like the ocean pressing on a deep-sea diver. A ton of rocks will crush the life out of you, but a ton of water pressure distributed over your entire body causes little discomfort (fig. 56).

Nitric acid formed beautiful crystals in the cell, while benzene, a liquid at room-pressure, revealed at least four different high-pressure crystal forms. By heating the diamond cell while he changed pressure, Van Valkenburg learned to change crystals from one form to another so he could map out the pressures and temperatures at which they are stable.

* * *

In spite of its power and simplicity, the diamond-anvil cell was slow to catch on (fig. 57). Most researchers who had spent their lives using impressive large-volume presses just couldn't get very excited about working on tiny volumes under a microscope. The point was underscored when Charles Weir made a special trip to New Hampshire, partly to show off the new device to the vacationing Percy Bridgman. Bridgman was politely interested, but hardly enthusiastic, and the visit was brief. Alvin Van Valkenburg received similarly neutral responses when he presented a series of lectures on the diamond-anvil cell. His low-key style and the prevailing use of big presses conspired against its immediate acceptance.

Of the three diamond cell inventors, Ellis Lippincott had the widest reputation. If the novel device was to gain new adherents, he was the one to promote it. But just as the diamond cell was coming into its own, Lippincott's reputation was destroyed by the polywater fiasco.

The scientific community was astounded in the late 1960s when Russian chemists N. N. Fedyakin and B. V. Derjaguin announced the

Fig. 57 Alvin Van Valkenburg examines a pressurized crystal in his diamond-anvil cell, *circa* 1964. (From the Author.)

discovery of a completely new kind of water that seemed to form easily in narrow capillary tubes.[11] The new water froze at $-50°C$, boiled at $300°C$, and appeared to be considerably more dense than ordinary water. The Russian researchers proposed that the new form of H_2O possesses a distinctive new atomic structure.

Ellis Lippincott and his co-worker, Robert R. Stromberg of the National Bureau of Standards, were skeptical but sufficiently intrigued by the unlikely Soviet reports to examine the new water themselves. They followed the Russian recipe and examined the minute sample with infrared spectroscopy – a sensitive probe of molecular structure – fully expecting to see the familiar water spectrum. Instead they stared in amazement at a totally new spectrum, unlike anything they'd seen before. This new form of water had all the traits of a polymer, a chain-like collection of atoms. Lippincott and his coworkers rushed their manuscript "Polywater," into print with a feature article in the June 27, 1969, issue of *Science*.[12]

For a brief time polywater was the rage, a source of intense study and wild speculation.[13] Scientists at the University of Southern California hinted at polywater's potential biological effects on skin pores – they called it a possible "fountain of youth." Frank Donahoe, a researcher at Wilkes College in Pennsylvania, gave readers of *Nature* a sober warning: "Polywater may or may not be the secret of Venus's missing water. The polymerization of Earth's water would turn her into a reasonable facsimile of Venus." A tiny amount of polywater released into the oceans, Donahoe argued, might cause all the water to transform. "I regard the polymer as the most dangerous material on earth.... Treat it as the most deadly virus until its safety is established." *The National Enquirer* picked up on the hype with the headline, "Scientists have discovered a new type of water that could poison the world!"

The conclusion to the sorry affair came at the hands of Dennis L. Rousseau of Bell Telephone Laboratories, who noted the similarity of the polywater spectrum to sodium lactate, a common ingredient in human perspiration. Rousseau wrote, "Determined to understand polywater's infrared spectrum, I turned to my athletic passion, handball. After a lively game, I returned to the laboratory with my sweaty T-shirt and wrung the perspiration into a flask. When I placed the sweat in an infrared spectrometer, the spectrum looked strikingly similar to that of polywater." A year and a half after the appearance of Lippincott's first article on polywater, *Science* published Rousseau's crushing rebuttal, "Polywater and sweat: Similarities between the infrared spectra."[14]

Polywater was Ellis Lippincott's last hurrah; a short time later he became ill with Hodgkin's disease and died in 1974. In spite of the ridicule caused by his intensely embarrassing error, Lippincott remained a gentleman. He could have exploited his initial findings to create personal publicity, but he avoided the circus-like atmosphere that has characterized the more recent cold fusion fiasco. He could have retaliated against his critics, but he refrained from antagonism or personal attacks. He just quietly slipped away, a sad and broken man.

* * *

Although scientists showed little interest in the Van Valkenburg, Weir, and Lippincott diamond cell until the late 1970s, the inventors knew they had a good idea. In 1960, the three scientists filed for a patent and formed a company to build and market the cells.[15] High Pressure Diamond Optics was founded in 1961 with an initial investment of $125.00 from each of the three co-inventors.[16]

Sales were slow in those first years, and only Van Valkenburg stayed with the business long enough to see its success. Charlie Weir retired in the late 1960s to engage in his favorite hobby, joining the color-blind world of amateur radio communications. Ellis Lippincott died in 1974.

The diamond-cell business took off in the mid 1970s for a reason few could have foreseen. Diamond cells can generate very high pressures, but that is not their only use. The opposed-anvil arrangement is also ideal for crushing soft samples into a uniform thickness ideal for spectroscopic identification. If you want to determine the origin of a chip of paint, a piece of mud, or perhaps a trace of white powder, the diamond cell is perfect for preparing the sample. Hence, the first major buyer of diamond cells from High Pressure Diamond Optics was the Royal Canadian Mounted Police. Today, forensics experts around the world use the device routinely in their investigations.

The company founded by Van Valkenburg, Weir, and Lippincott is now a thriving concern that has sold more than a thousand diamond cells. Their main business remains forensic devices, but High Pressure Diamond Optics also offers state-of-the-art lever arm models, modifications of the original NBS pressure cell that achieve record high pressures of several million atmospheres.

In his final years, Alvin Van Valkenburg passed the daily business responsibilities to his son Eric, but still retained his unbridled enthusiasm for the diamond anvil cell. He loved to travel to conferences and conventions, where he'd set up a diamond-cell display to show high-pressure phenomena to anyone who wanted to peer down his microscope. He would delight in retelling the story of his first glimpse of the high-pressure world, and the sweep of his many subsequent discoveries.

And he would say, over and over again, "We had such a wonderful time!"

Notes

1. C. D. Clark, P. J. Dean, and P. V. Harris, "Intrinsic edge absorption in diamond." *Proceedings of the Royal Society, London* **A277**, 312–329 (1964). See also Gordon Davies, *Diamond*. (Bristol: Adam Hilger, 255 pp. 1984).
2. A. W. Lawson, First through Fourth "Annual report to ONR on high-pressure research (ONR Contract No. N-6ori-20–XX)." 1947–1960.
3. A. W. Lawson and T.-Y. Tang, "A diamond bomb for obtaining powder pictures at high pressures". *Review of Scientific Instruments* **21**, 815–817 (1950). Recollections of Andrew Lawson are found in: Julian R. Goldsmith, "Some Chicago georecollections. *Annual Review of Earth and Planetary Sciences* **19**, 1–16 (1991). Additional insight was provided by interviews with Hatten S. Yoder, Jr.
4. John C. Jamieson, "Introductory studies of high-pressure polymorphism to 24,000 bars by x-ray diffraction with some comments on calcite II." *Journal of Geology* **65**, 334–343 (1957).
5. The history of the diamond anvil cell was obtained primarily through interviews with Alvin Van Valkenburg. Additional information was provided during interviews by his son, Eric Van Valkenburg, and with Julian Goldsmith, John Jamieson, Gaspar Piermarini, and Hatten S. Yoder, Jr.
6. For a more complete discussion of diamond types, see: George E. Harlow (editor), *The Nature of Diamond*. (Cambridge: Cambridge University Press, 278 pp. 1998).
7. The National Bureau of Science drilled diamond cell was described in a press release of December 22, 1957: "Confiscated diamond serves NBS scientists as 'test tube' for high pressure studies." See also "Know your bureau." *The NBS Standard* **3**, 1–2 (December, 1957).
8. J. C. Jamieson, A. W. Lawson, and N. D. Nachtrieb, "New device for obtaining x-ray diffraction patterns from substances exposed to high pressure." *Review of Scientific Instruments* **30**, 1016–1019 (1959).
9. C. E. Weir, E. R. Lippincott, A. Van Valkenburg, and E. N. Bunting, "Infrared studies in the 1– to 15–micron region to 30 000 atmospheres. *Journal of Research of the National Bureau of Standards* **63A**, 55–62 (1959). The first illustrations of high-pressure transitions appear in: A. Van Valkenburg, "Visual observations of high-pressure transitions." *Review of Scientific Instruments* **33**, 1462 (1962).

10. The metal gasketing technique is described by: A. Van Valkenburg, "Diamond high-pressure windows." *Diamond Research* 17–19 (1964). Color illustrations of "gincicles" appeared in: Joyce Schroeder, "Through a glass, brightly." The Washington Post Potomac, December 29, 1963.

11. Boris V. Derjagen and N. V. Churaev, "Nature of 'anomalous water'." *Nature* 244, 430–431 (1973).

12. Ellis R. Lippincott, Robert R. Stromberg, Warren H. Grant, and Gerald L. Cessac, "Polywater: Vibrational spectra indicate unique stable polymeric structure." *Science* 164, 1482–1487 (1969).

13. F. J. Donahoe, "Anomalous water." *Nature* 224, 198 (1969).

14. Denis L. Rousseau "'Polywater' and sweat: similarities between the infrared spectra." *Science* 171, 170–172 (1971); see also: Denis L. Rousseau and Sergio P. S. Porto, "Polywater: polymer or artifact?" *Science* 167, 1715–1719 (1970); and Denis L. Rousseau, "Case studies in pathological science." *American Scientist* 80, 54–63 (1992). Additional details of the polywater episode were provided in interviews with Denis Rousseau and Alvin Van Valkenburg.

15. C. E. Weir, A. Van Valkenburg, and E. R. Lippincott, US Patent No. 3,079,505, "High-pressure optical cell," filed August 26, 1960, granted February 26, 1963. A modified version of his cell was described in A. Van Valkenburg, E. R. Lippincott, and C. E. Weir, US Patent No. 3,560,091, "High-pressure optical cell," filed April 12, 1968, granted February 2, 1971.

16. Alvin Van Valkenburg and Eric Van Valkenburg provided details of the history of High-Pressure Diamond Optics, Inc.

EPILOGUE

S O MUCH WORK REMAINS TO BE DONE. High-pressure research now consumes the lives of thousands of scientists and engineers, who know there are many breakthroughs still to be made. For some researchers the quest for exotic new materials, such as metallic hydrogen, high-temperature superconductors, and exceptional abrasives, stands as the central problem in high-pressure science. Others investigate the behavior of high-pressure minerals that are presumed to form the bulk of the solid earth's deep interior. And a growing legion of physicists employ pressure to probe the curious properties of matter under extreme conditions.

In the midst of this intensive research effort, much fame and fortune still awaits the creative diamond maker. Countless investigators will spend years trying to figure out the best ways to grow isotopically pure diamond, to synthesize diamonds by explosion, and to form semiconducting diamonds by vapor deposition. The diamond gemstone industry is not about to collapse, and in all likelihood there will always be a demand for fine natural stones, but as diamond abrasives, thin films, and other objects become more and more available to industry and commerce, they will play an ever-increasing role in our lives. Products that rely on futuristic high-pressure technologies and materials will insinuate themselves into every aspect of our lives: cheaper manufacturing, faster transportation, more reliable communications, stronger construction, and new forms of recreation.

The high-pressure adventure of the past century has featured remarkable transformations in our technical capabilities and in our understanding of matter. In the decades since Percy Bridgman first broke the 2000–atmosphere mark in a makeshift laboratory, the range of routine research pressures has increased a thousand-fold, and new records are set every few years. Under these extremes scientists have discovered countless new materials. Yet we have only just begun to discover the astonishing variety of exotic compounds and unexpected physical phenomena that pressure can provide. What new wonders await us? What unimagined substances will change our lives?

High-pressure scientists are ever aware of these extraordinary possibilities, yet we are captivated by pressure not so much by thoughts of profit, but rather by nature's endless store of magic. Each day's trip to the laboratory begins an exploration into uncharted realms – every day holds the prospect of discovering fantastic materials and observing strange phenomena that no one has ever seen before.

For us, science remains the greatest adventure of all.

INDEX

CPSIA information can be obtained
at www.ICGtesting.com
Printed in the USA
LVHW032042040220
645827LV00004B/598